EARLY DAYS OF OIL

A PICTORIAL HISTORY OF THE BEGINNINGS

OF THE INDUSTRY IN PENNSYLVANIA

EARLY DAYS OF
OIL

A PICTORIAL HISTORY OF THE BEGINNINGS OF THE INDUSTRY IN PENNSYLVANIA

BY PAUL H. GIDDENS

PRINCETON, NEW JERSEY
PRINCETON UNIVERSITY PRESS · 1948

London: Geoffrey Cumberlege, Oxford University Press

Printed in the United States of America by Meriden Gravure Co.
Composed by Princeton University Press

To

JOHN A. MATHER

Through whose untiring effort a priceless pictorial record of the beginning of the petroleum industry was made and preserved

PREFACE

THE purpose of this volume is to present chronologically a pictorial history of the early days of the petroleum industry. Pictures of famous wells, farms, teamsters, river transportation facilities, pipe lines, shooting wells, tank cars, engines and boilers, refineries and many other things naturally form an essential part of the record. Other pictures, however, have been included to illustrate life in the oil region—how people lived and dressed, the churches and schools they attended, the kind of towns in which they lived, the amusements they enjoyed and the dangers to which they were exposed. Except for the introductory statement at the beginning of each chapter and the brief descriptive notes accompanying the pictures, the narrative has been deliberately kept to a minimum.

Thanks to John A. Mather, probably no industry has as fine or as extensive a pictorial record of its beginnings as the petroleum industry. It is impossible to tell how many pictures Mather made. During the fire and flood in Titusville in June, 1892, he lost more than 16,000 negatives but over 3,000 were saved. After Mather's death in 1915 these negatives were stored in different places. When the Drake Well Memorial Park near Titusville was established in 1934, his collection was placed in the Drake Museum where it is to be found today. These old wet-plate glass negatives are in excellent condition for the most part. The collection constitutes a priceless pictorial record of one of the most exciting industrial eras in American history.

Most of the photographs in this volume have been taken directly from the original Mather negatives. Where a negative has been broken or destroyed, pictures have been taken from existing prints. Of the 3,274 Mather negatives in the Drake Museum, 2,229 have been partially or wholly identified and labeled. Every effort has been made to provide accurate and reliable identification of the pictures used in this book but, in some cases, it has been exceedingly difficult owing to the lack of adequate records. For this reason the author would welcome correspondence from anyone who could assist in a better identification. In order to provide as complete a story as possible, additional pictures and illustrations have been drawn from various contemporary sources.

Since Mather lived in Titusville and most of the oil developments from 1859 to 1870 took place within a radius of about twenty-five miles, most of the pictures in the Mather collection relate to this region. As oil operations moved farther away—down the Allegheny River and then up to Bradford—Mather could not follow the oilmen as easily as during the first decade. Otherwise, the pictorial history of these areas would have been more adequately presented.

In preparing the narrative the author has drawn heavily and without using quotes from his book, *The Birth of the Oil Industry*, which has been done by permission of the publisher, The Macmillan Company. For pictures taken from books, magazines, and newspapers, the author wishes to make special acknowledgment. A complete list of all pictures and the source from which they have been taken will be found at the end of the book.

I am particularly indebted and grateful to W. C. Wenzel, Executive Manager, and the Board of Directors of the Pennsylvania Grade Crude Oil Association, Oil City, for their encouragement and material assistance in the preparation of this volume; to the Drake Museum, Titusville, for the privilege and opportunity to use the Mather collection and other pictures; and to Datus C. Smith, Jr., Director, P. J. Conkwright, Typographer, and Helen Van Zandt of the Princeton University Press, with whom it has been a distinct pleasure to work in preparing this book for publication.

PAUL H. GIDDENS

Allegheny College
Meadville, Pennsylvania

CONTENTS

	Preface	v
I.	The Beginning	1
II.	Drilling Along Oil Creek	9
III.	Pioneer Towns Along Oil Creek	36
IV.	Transporting Oil From Oil Creek	53
V.	Pithole	60
VI.	Revolutionizing the Production and Transportation of Oil	66
VII.	Between Oil Creek and the Allegheny River	76
VIII.	Titusville, the Queen City	89
IX.	Oil Developments 1870 to 1885	103
X.	Great Oil Fires	125
XI.	Honoring Drake and the Birth of the Petroleum Industry	136
	List of Photographs	145
	Index	148

CHAPTER I

THE BEGINNING

LONG before Edwin L. Drake drilled his famous oil well near Titusville, Pennsylvania, petroleum was known to exist and was used in the United States. Seventeenth-century French missionaries allude in their journals to oil in western New York. In the eighteenth century there are reports of a trade in oil brought to Niagara by the Seneca Indians; this probably gave rise to the early name "Seneca Oil" for petroleum. Prior to 1846, however, the greatest source of petroleum in the United States was to be found along Oil Creek in northwestern Pennsylvania. As white settlers moved into this region after the American Revolution and settled along the Creek, they began to skim petroleum from little springs either in the bank or in the actual bed of the stream. They valued and used petroleum exclusively as medicine.

Petroleum was not used in great quantities nor for commercial purposes until about 1847 when Samuel M. Kier of Pittsburgh began bottling and selling petroleum as medicine from his father's salt wells near Tarentum, Pennsylvania. Despite its low price, Kier could not dispose of all the oil produced by these wells. Having burned crude oil at the Tarentum wells, Kier believed he might use the surplus only if some method could be found to eliminate the smoke and odor. After much experimentation Kier devised a crude distillation process, and about 1850 he began to distill petroleum, calling the new product "carbon oil." Since it was cheaper, safer and better than any existing illuminant, "carbon oil" came into general use in western Pennsylvania and a thriving trade developed in New York City. The demand soon exceeded the supply; the price jumped from seventy-five cents a gallon to $1.50 and then to $2.00. All efforts to increase the supply met with indifferent success until the drilling of the Drake well in August 1859, when Drake solved the perplexing problem and demonstrated petroleum could be secured in sufficient quantities to market it commercially. This epoch-making event marked the launching of the petroleum industry.

EARLIEST RECORD OF PETROLEUM IN PENNSYLVANIA. Lewis Evans' *Map of the Middle British Colonies in America*, published in 1755, is the first record to indicate the presence of petroleum in Pennsylvania. Note that the word "Petroleum" is printed very close to the present site of Oil City.

FIRST PETROLEUM SHIPPED TO PITTSBURGH. About 1790 Nathaniel Carey, one of the first settlers on Oil Creek in northwestern Pennsylvania, began collecting oil from the springs and seepages along Oil Creek and peddling it through the country. Carey is said to have introduced petroleum in Pittsburgh.

4 Axes
1 Grubing Hoa
1 Iron Wedge
1 Socket
1 Small drawing Knife
1 Bell
1 Round Shave
1 Bake Kettle
1 Skillet
3 Kegs Senica Oil 50 Dllrs
3 bbl Ion bound & Spirit
6 D. Old
4 Old Rifles
2 Houses without

EARLY QUOTATION ON OIL, 1797. General William Wilson kept a general store at Fort Franklin at the junction of French Creek and the Allegheny River. In his Day Book for 1797 an inventory of goods shows "3 Kegs Senica Oil 50 Dllrs," which is one of the earliest, if not the earliest, records on the price of petroleum.

COLLECTING OIL ON OIL CREEK, 1810. In 1810 J. Francis Waldo made a sketch of some men skimming petroleum from an oil spring on Oil Creek near the present site of Titusville. The oil is being placed in hollowed-out-logs on a raft to be floated down the Creek and the Allegheny River to Pittsburgh.

SAMUEL M. KIER

Impressed by the medicinal value of petroleum, Kier, an owner and operator of canalboats between Pittsburgh and Philadelphia, opened an establishment in Pittsburgh about 1847 where petroleum was put up in half-pint bottles. Through agents who traveled about the country, petroleum was sold to the public as a cure for all ailments, human or animal.

Although Kier widely publicized petroleum as a medicine, his supply of petroleum exceeded the demand, so he concluded that something leading to a more general utilization of oil must be done. After consulting a prominent Philadelphia chemist, Kier became convinced that by distilling petroleum he could obtain an excellent illuminant. Immediately he erected a one-barrel still on Seventh Avenue above Grant Street in Pittsburgh. Here, about 1850, he began to distill petroleum and became America's pioneer oil refiner. The demand for Kier's "carbon oil" was so great that he soon had to install a five-barrel still.

Kier's Petroleum, or Rock Oil, and the wrapper in which it was sold.

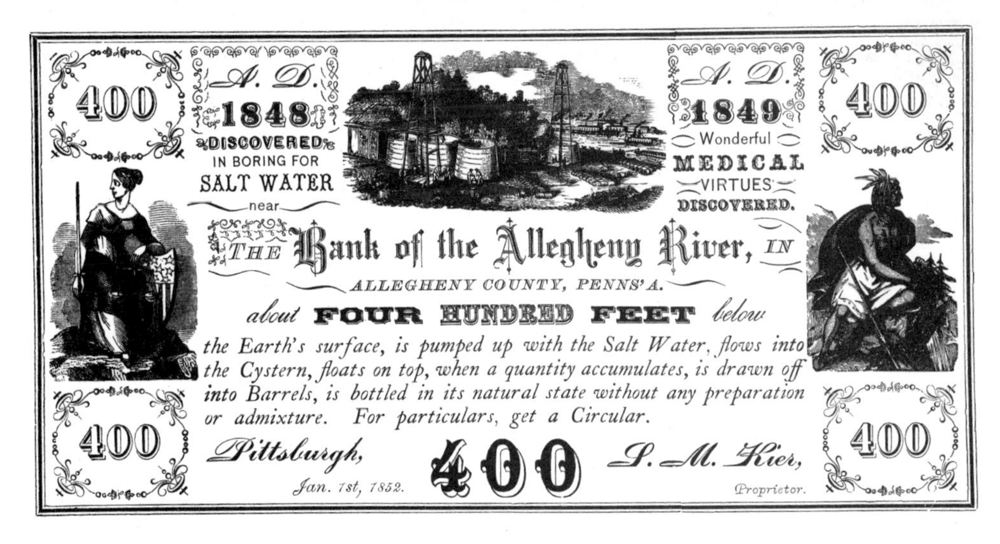

Kier's Advertisement and Circular, describing the wonderful curative properties of Kier's petroleum.

PETROLEUM, OR ROCK OIL.

A NATURAL REMEDY!

PROCURED FROM A WELL IN ALLEGHENY COUNTY, PA.

Four hundred feet below the Earth's Surface!

PUT UP AND SOLD BY

SAMUEL M. KIER,

CANAL BASIN, SEVENTH STREET, PITTSBURGH, PA.

The healthful balm from Nature's secret spring,
The bloom of health, and life, to man will bring;
As from her depths the magic liquid flows,
To calm our sufferings, and assuage our woes.

CAUTION.—As many persons are now going about and vending an article of a spurious character, calling it Petroleum, or Rock Oil, we would caution the public against all preparations bearing that name not having the name of S. M. KIER written on the label of the bottle.

PETROLEUM.—It is necessary, upon the introduction of a new medicine to the notice of the public, that something should be said in relation to its powers in healing disease, and the manner in which it acts. Man's organization is a complicated one; and to understand the functions of each organ, requires the study of years. But to understand that certain remedies produce certain impressions upon these organs, may be learned by experience in a short time. It is by observation in watching the effects of various medicines, that we are enabled to increase the number of curative agents; and when we have discovered a new medicine and attested its merits, it is our duty to bring it before the public, so that the benefits to be derived from it may be more generally diffused, but have no right to hold back a remedy whose powers are calculated to remove pain and to alleviate human suffering and disease. THE PETROLEUM HAS BEEN FULLY TESTED! About one year ago, it was placed before the public as A REMEDY OF WONDERFUL EFFICACY. Every one not acquainted with its virtues, doubted its healing properties. The cry of humbug was raised against it. It had some friends;—those that were cured through its wonderful agency. These spoke out in its favor. The lame, through its instrumentality, were made to walk—the blind, to see. Those who had suffered for years under the torturing pains of RHEUMATISM, GOUT and NEURALGIA, were restored to health and usefulness. Several who were blind have been made to see, the evidence of which will be placed before you. If you still have doubts, go and ask those who have been cured! Some of them live in our midst, and can answer for themselves. In writing about a medicine, we are aware that we should write TRUTH—that we should make no statements that cannot be proved. We have the witnesses—crowds of them, who will testify in terms stronger than we can write them to the efficacy of this Remedy, who will testify that the PETROLEUM has done for them what no medicine ever could before—cases that were pronounced hopeless, and beyond the reach of remediate means—cases abandoned by Physicians of unquestioned celebrity, have been made to exclaim, "THIS IS THE MOST WONDERFUL REMEDY EVER DISCOVERED!" We will lay before you the certificates of some of the most remarkable cases; to give them all, would require more space than would be allowed by this circular. Since the introduction of the Petroleum, about one year ago, many Physicians have been convinced of its efficacy, and now recommend it in their practice; and we have no doubt that in another year it will stand at the head of the list of valuable Remedies. If the Physicians do not recommend it, the people will have it of themselves—for its transcendent power to heal, *will* and *must* become known and appreciated—when the voices of the cured speak out; when the cures themselves stand out in bold relief, and when he who for years has suffered with the tortures and pangs of an immedicable lesion, that has been shortening his days, and hastening him "to the narrow house appointed for all the living," when he speaks out in its praise, who will doubt it? THE PETROLEUM IS A NATURAL REMEDY—it is put up as it flows from the bosom of the earth, without anything being added to or taken from it.

FRANCIS BEATTIE BREWER

GEORGE H. BISSELL

REPORT

ON THE

ROCK OIL, OR PETROLEUM,

FROM

VENANGO CO., PENNSYLVANIA.

WITH SPECIAL REFERENCE TO ITS USE FOR ILLUMINATION AND OTHER PURPOSES.

BY B. SILLIMAN, JR.,

PROF. OF GENERAL AND APPLIED CHEMISTRY, YALE COLLEGE.

COPYRIGHT SECURED.

NEW HAVEN:

FROM J. H. BENHAM'S STEAM POWER PRESS.

1855.

SILLIMAN'S REPORT

In the fall of 1853 Francis Beattie Brewer, a graduate of Dartmouth College and a physician in Titusville, Pennsylvania, carried a small bottle of petroleum on a trip to Hanover, New Hampshire to visit relatives and friends. The sample of petroleum had been taken from an oil spring on the farm of Brewer, Watson & Company about two miles south of Titusville. At Dartmouth Dr. Dixi Crosby and Professor O. P. Hubbard examined the oil and pronounced it very valuable.

A few weeks later George H. Bissell, another Dartmouth graduate and a young lawyer in New York City, returned to his home in Hanover, saw the bottle of petroleum in Crosby's office, and immediately became interested in its commercial possibilities for illuminating purposes. As a result, Bissell and his partner, Jonathan G. Eveleth, in November 1854, bought the farm with the oil springs from Brewer, Watson & Company, organized the Pennsylvania Rock Oil Company of New York on December 30, 1854, and prepared to secure petroleum in large enough quantities to put on the market.

In order to determine the economic value of petroleum and make it easier to sell stock in the new oil company, Eveleth and Bissell engaged one of the most distinguished scientists of the day, Professor Benjamin Silliman, Jr., of Yale College, to analyze the oil.

BENJAMIN SILLIMAN, JR.

Completed in April 1855, Silliman's analysis proved to be a decisive factor in the establishment of the petroleum industry, for it not only dispelled many doubts about petroleum but induced capitalists to invest in the enterprise.

Among the capitalists interested in the venture of Eveleth nd Bissell was James M. Townsend, President of the City avings Bank of New Haven, Connecticut. He and some of is associates induced Eveleth and Bissell to abandon the New York company and incorporate in Connecticut where he property of a stockholder was not liable for the debts of company as in New York. The Pennsylvania Rock Oil Company of Connecticut, therefore, came into existence on September 18, 1855, and within a short time all the capital had een subscribed, mostly by New Haven men.

Owing to a lack of harmony which unexpectedly developed between the New Haven stockholders and Eveleth and Bissell, Townsend and his associates organized the Seneca Oil Company of Connecticut on March 23, 1858. Then, as majority stockholders of the Pennsylvania Rock Oil Company of Connecticut, they leased the oil farm to themselves s stockholders of the Seneca Oil Company.

JAMES M. TOWNSEND

Seneca Oil Company, New Haven, Ct.

500,000

No.

SHARES

CAPITAL $500,000 STOCK.

This is to Certify that

is entitled to Shares of the Capital Stock of the SENECA OIL COMPANY transferable only on the Books of the Company by or Attorney upon the surrender of this Certificate.

IN TESTIMONY WHEREOF, the said Company have caused this Certificate to be signed by their President and Secretary

500,000

SHARES $25 EACH

Sec.y Pres.t

EDWIN L. DRAKE

Born in 1819, Drake spent the early years of his life on a farm first in New York and later in Vermont. With only a common school education, Drake left home at the age of nineteen and became a jack-of-all-trades. In 1849 he worked as a conductor on the New York & New Haven Railroad and lived in New Haven. While living here, he became acquainted with Townsend, who persuaded him to buy some oil stock. During the summer of 1857 Drake fell ill and was forced to give up his work with the railroad. Since he was idle and could obtain a railroad pass, Townsend sent him to Titusville to examine the oil farm. On the basis of Drake's report Townsend organized the Seneca Oil Company. The stockholders appointed Drake General Agent of the company at an annual salary of $1,000 and sent him to Titusville to drill for oil.

On the site of the principal c
spring of the Brewer, Watson & Com
pany farm, Drake built an engin
house, erected a derrick in which
swing the drilling tools, and installe
his engine and boiler. An iron pi
was driven 32 feet through the quic
sands and clay into bedrock. The dri
ing tools were placed inside the pip
and about the middle of August 185
they began to drill, averaging abo
three feet a day. On Saturday afte
noon, August 27, just as Smith and th
workmen were about to quit for th
day, the drill dropped into a crevi
at 69 feet and slipped down six inche
The men pulled out the tools and we
home. Late Sunday afternoon "Unc
Billy," as Smith was affectionate
called, visited the well, peered into th
pipe, and saw oil floating on top of th
water within a few feet of the derri
floor. They had struck oil! Drake ha
demonstrated how oil could be s
cured in greater abundance. He ha
tapped a vast subterranean deposit
petroleum and thus ushered in a ne
industry—the petroleum industry.

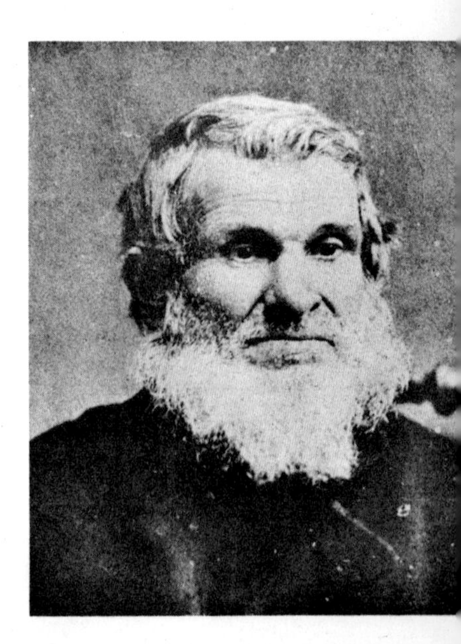

WILLIAM A. SMITH

A blacksmith and experienced sa
well driller from Tarentum employ
by Drake to drill for oil at $2.50 a da

The Drake Well in 1861

Drake stands in front of his well with Peter Wilson, a Titusville druggist.

TYPE OF ENGINE AND BOILER
USED BY DRAKE

To furnish power for drilling, Drake used a six-horsepower engine and a "Long John" stationary, tubular boiler.

DRAKE'S DRILLING TOOLS

Drake's drilling tools were made by William A. Smith a Tarentum. They weighed 100 pounds and cost $76.50.

THE DRAKE WELL IN 1864. "Uncle Billy" Smith is sitting on the wheelbarrow in the foreground. The smaller of the two girls, Annette Farwell, now Mrs. Samuel Grumbine, still lives in Titusville.

CHAPTER II

DRILLING ALONG OIL CREEK

ONSIDERING the communication facilities of the time, the apid spread of the news of Drake's discovery was phenom-nal. It seemed as if the entire population of Titusville and he surrounding country had heard the news simultaneously, or in less than twenty-four hours, hundreds of people were nilling around the Drake well. Everyone was wild to lease r buy land at any price and drill a well. Because of the loca-ion of Drake's well, it was believed that the best place to rill was in the lowlands and as near as possible to the vater. Consequently, there was a mad rush to secure land ear the Drake well and along the Creek. Land bordering on il Creek was soon taken up and within a relatively short ime, the entire valley as far back as and even into the hill-ides, had been leased or purchased. Soon scores of wells vere being drilled.

LOWER OIL CREEK & VICINITY

SCALE 0 1 2 MILES

Except for a few early wells, like the Barnsdall, Crossley, Williams and others, which were drilled near the Drake well, the main scene of drilling operations between 1860 and 1868 was in the lower portion of Oil Creek valley. This district produced the flowing wells. It had the best producing farms, and in that period, produced about two-thirds of all the oil brought to the surface.

The pioneer wells of 1859 and 1860 produced more oil than anyone had ever seen and created fierce excitement throughout the region but they were nothing compared to the great flowing wells, like the Empire, Sherman, Phillips, Woodford, the Noble and Delamater, and others along Oil Creek beginning in 1861. They flowed from 1,500 to 4,000 barrels a day. At the end of 1860, there were all together about seventy-four producing wells and they yielded daily about 1,200 barrels. In 1864, as a result of flowing wells, the daily production had jumped to about 6,000 barrels. Oil Creek valley was a bee-hive of activity.

LOCATING AN OIL WELL, WITCH HAZEL METHOD. Guessing tricks and superstitious devices for locating a place to drill flourished everywhere. Many of the operators used the divining rod. With a Y-shaped witch hazel stick held firmly in both hands in a horizontal position, a man would walk slowly about the plot. If the loose end happened to be pulled toward the ground by an unseen force, this was the spot on which to sink a well. Other owners relied more heavily upon dreams to locate a site. Still others, with a more prosaic turn of mind, depended upon the sense of smell to lead them to a suitable spot.

KICKING DOWN A WELL. Most of the early wells were drilled by a slow but simple process called "kicking down a well," or the spring pole method. An elastic pole, about fifteen feet long, was placed over a fulcrum with the large end fastened to the ground. Two or three feet from the free end, the drilling tools were connected with the pole and dropped into the driving pipe. Attached to the free end of the pole were stirrups in which two men each placed a foot and pulled down, permitting the tools to drop on the rock. When they loosened their hold, the spring of the pole pulled it back with enough force to raise the drilling tools a few inches. The same thing could be done by using a platform, as above. Repeating this procedure rapidly all day long enabled them to drill on an average about three feet a day. Though laborious, the spring pole method provided men of moderate means and strong muscles with a cheap method for sinking a well in shallow territory.

a—ROCK STRATA.
b—EARTH'S CRUST.
c—SAMSON POST.
d—DERRICK.
e—BULL WHEEL.
f—WALKING BEAM.
g—TEMPER SCREW.
h—ROPE.
i—ROPE SOCKET.
k—JARS.
l—AUGUR STEM.
m—CENTRE-BIT.

This sketch shows the common method which was used to drill wells once the spring pole method had been abandoned.

Those who could afford engines and boilers used steam power to drill. The picture shows one type of engine and boiler used in the sixties.

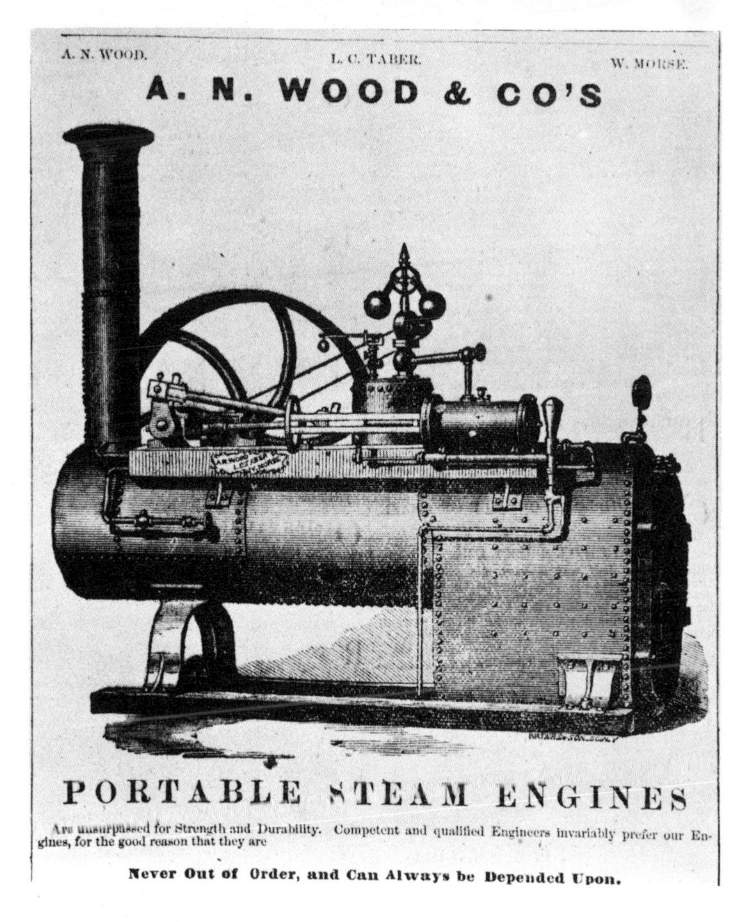

Inside View of a Derrick

THE EMPIRE WELL, FUNK FARM

The first great flowing well was the Empire. Completed in September 1861, it started flowing 3,000 barrels a day! Unable to secure barrels at any price, the owners tried to check the flow of oil but without success. They built a dam around the well and let it run into the enclosure; but the oil refused to be confined and ran off into Oil Creek. The yield bewildered the owners. It was too good a thing. With the market already flooded and some 3,000 more barrels added to the daily supply, the Empire drove the price of petroleum down to ten cents a barrel.

Within a few weeks after the Empire started flowing, it was eclipsed by a new well on the Tarr farm at the lower end of Oil Creek. In October 1861, the Phillips well began flowing 4,000 barrels; it finally settled down to 2,500 barrels and continued at this rate for months. About four rods away from the Phillips, N. S. Woodford drilled another well, which flowed 1,500 barrels in July 1862. In the picture on the opposite page, the Phillips well is to the right and the Woodford to the left.

Phillips and Woodford Wells, Tarr Farm

Underground Storage Tanks, Tarr Farm

Completely unprepared for the flood of oil which came from the Phillips and Woodford, the owners dug holes in the ground and cribbed them with timber in order to store the oil. In time, these tanks covered several acres.

THE SHERMAN WELL, FOSTER FARM

Probably the best single strike in 1862 was the Sherman well. Its owner, J. W. Sherman of Cleveland, came to the oil region in the early days and, like many others, was a man of limited means. Securing a lease, he sank a well with a spring pole. In March 1862, he tapped a vein of oil which spouted 1,500 barrels a day. It averaged about 900 barrels daily for two years and it brought the owner about $1,700,000. Mather himself had an opportunity to buy an eighth interest in this well for $68 but he failed to seize the opportunity. This investment would have netted him $175,000.

THE NOBLE AND DELAMATER WELL, FARREL FARM

Renowned for the Phillips and Voodford wells, the Tarr farm of two undred acres was one of the best pro-ucing farms along the Creek. The mazing output of the Phillips and Voodford created a terrific demand or leases, and scores of wells were rilled. James Tarr sold his interest in ne farm in 1865 for gold equivalent to 2,000,000 in currency and moved to Ieadville. Another $1,000,000 would ardly cover his royalty income.

The best paying well in the region vas drilled by Orange Noble and George B. Delamater, merchants from 'ownville, Pennsylvania. While they vere drilling with a spring pole one ay in May 1863, oil suddenly burst orth, rising over one hundred feet in he air. It flowed at the rate of 3,000 arrels a day. For days the oil ran into Dil Creek but finally it was brought nder control. With oil at $4.00 a bar-el and the price steadily rising, Noble nd Delamater's daily income varied rom $12,000 to $45,000. All together he well produced over 1,500,000 bar-els of oil and netted the owners over 5,000,000.

A Women's Party, Tarr Farm

BARREL YARD, SHAFFER FARM. Neither Colonel Drake nor he other pioneer oil operators were prepared for the flood f oil produced by the early wells. In 1859 Drake used wash ubs and boilers and what empty whiskey barrels he could ecure in Meadville and Titusville to hold his oil. Some of he other producers dug holes in the ground, cribbed them vith timbers, and let the oil run into these holes.

Loading Platform, Shaffer Farm

"Fishing" for Lost Drilling Tools, Oil Creek

OIL DIPPERS AT MILLER FARM, 1863. So much oil ran to wast down Oil Creek because of flowing wells or leaky bulk boat that people along the Creek would stand on the banks wit buckets and skim oil off the top of the water, then pour i into barrels. Many a man got started in the oil business i this way.

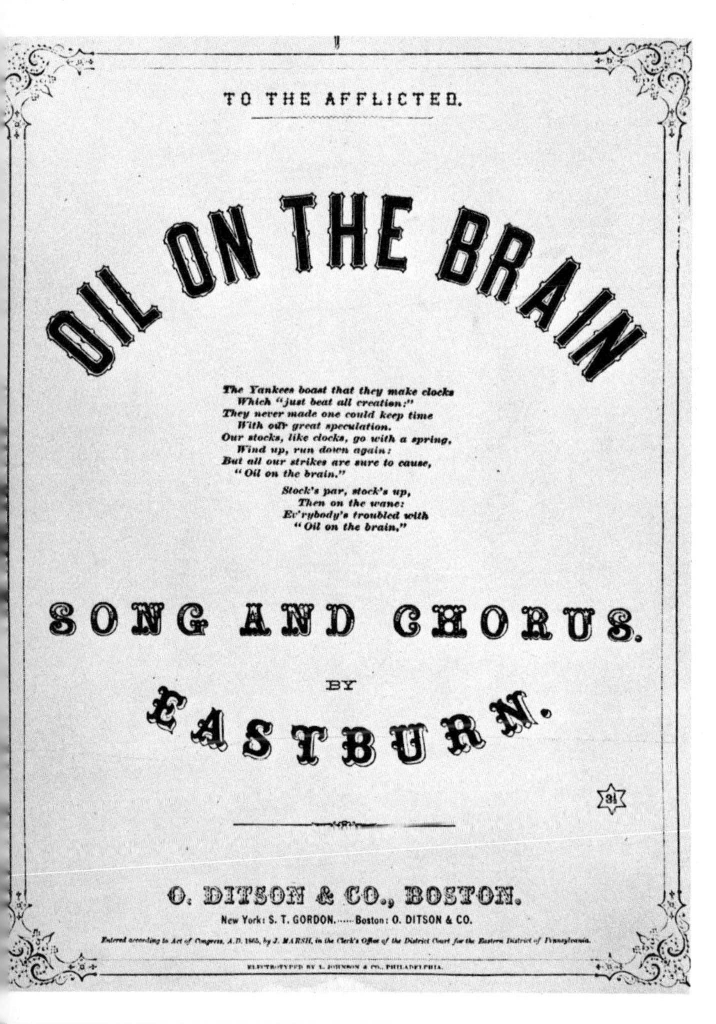

BARREL FACTORY, OIL CREEK. Within a short time cooperage shops came into existence everywhere throughout the oil region to supply barrels. In 1860 they sold for $2.00 apiece and the demand exceeded $1,000 a day. Coopers could not keep pace with the demand and the daily loss of oil was appalling.

The excitement created by the oil development along Oil Creek between 1859 and 1864 was reflected in popular songs and instrumental music. At least a half dozen pieces of sheet music about petroleum appeared in 1865.

OIL ON THE BRAIN

The Yankees boast that they make clocks Which "just beat all creation:"
They never made one could *keep time* With our great speculation.
Our stocks, like clocks, *go with a spring*, Wind up, run down again:
But all our *strikes* are sure to cause "Oil on the brain."

Chorus: Stock's par, stock's up, Then on the wane:
Everybody's troubled with "Oil on the brain."

There's various kinds of oil afloat, Cod-liver, Castor, Sweet;
Which tend to make a sick man *well*, And set him on his *feet*.
But ours a curious *feat* performs: We just a *well* obtain,
And set the people crazy with "Oil on the brain."

There's neighbor Smith, a poor young man, Who couldn't raise a dime;
Had clothes which boasted many rents, And took his "nip" on time.
But now he's clad in dandy style, Sports diamonds, kids, and cane;
And his success was owing to "Oil on the brain."

Miss Simple drives her coach-and-four, And dresses in high style;
And Mister Shoddy courts her strong, Because her "Dad's struck ile."
Her jewels, laces, velvets, silks, Of which she is so vain,
Were bought by "Dad" the time he had "Oil on the brain."

You meet a friend upon the street; He greets you with a smile,
And tells you, in a hurried way, He's "just gone into ile."
He buttonholes you half an hour: Of course you can't complain,
For you can see the fellow has "Oil on the brain."

The lawyers, doctors, hatters, clerks, Industrious and lazy,
Have put their money all in stocks, In fact, have gone "oil crazy."
They'd better stick to briefs and pills, Hot irons, ink, and pen;
Or they will "kick the bucket" from "Oil on the brain."

Poor Mrs. Jones was taken ill: The doctors gave her up.
They lost the confidence they had In Lancet, leech, and cup.
"Affliction sore long time she bore; Physicians were in vain;
And at last expired of "Oil on the brain."

There's "Maple Shade," "Excelsior," "Bull Creek," "Big Tank," "Dalzell,"
And "Keystone," "Star," "Venango," "Briggs," "Organic," and "Farrell,"
"Petroleum," "Saint Nicholas," "Corn Planter," "New Creek Vein."
Sure, 'tis no wonder many have "Oil on the brain."

Cow Run Wells. Cow Run and Bull Run were tributaries of Oil Creek near Shaffer Farm.

BULL RUN WELLS

MAGILL WELLS, FOSTER FARM

PIONEER RUN WELLS

An Oilman's Home, Pioneer Run

Pioneer Run was almost unknown as an oil producing territory until the completion of the Lady Brooks well in April 1866, which flowed over 600 barrels a day. Speculators eagerly sought the side hills; operators began putting down other wells not only on Pioneer Run but on Western Run, a tributary stream. Daily production soon amounted to 2,500 barrels.

GREAT WESTERN RUN WELLS

MONITOR REFINERY, G. W. McCLINTOCK FARM, 1864

The early refineries along Oil Creek were of this crude and primitive type. The number of refineries in the oil region rapidly increased. At the end of 1860 fifteen had been established, all of them of small capacity. Three years later, sixty-one refineries, with capacities varying from fifteen to three hundred barrels a day, dotted the oil region.

PETROLEUM CENTRE

The Stevenson farm lay on Benninghoff Run. In 1865 the Phillips brothers drilled a well on an elevation of about 250 feet above Oil Creek. Late in August, the Ocean well, as it was called, began flowing 300 barrels a day. This well did much to explode the theory that producing wells could not be found on hills, and the immediate result was to attract operators to the area. In the succeeding months, the Stevenson, Benninghoff, and Boyd farms vied with one another in production.

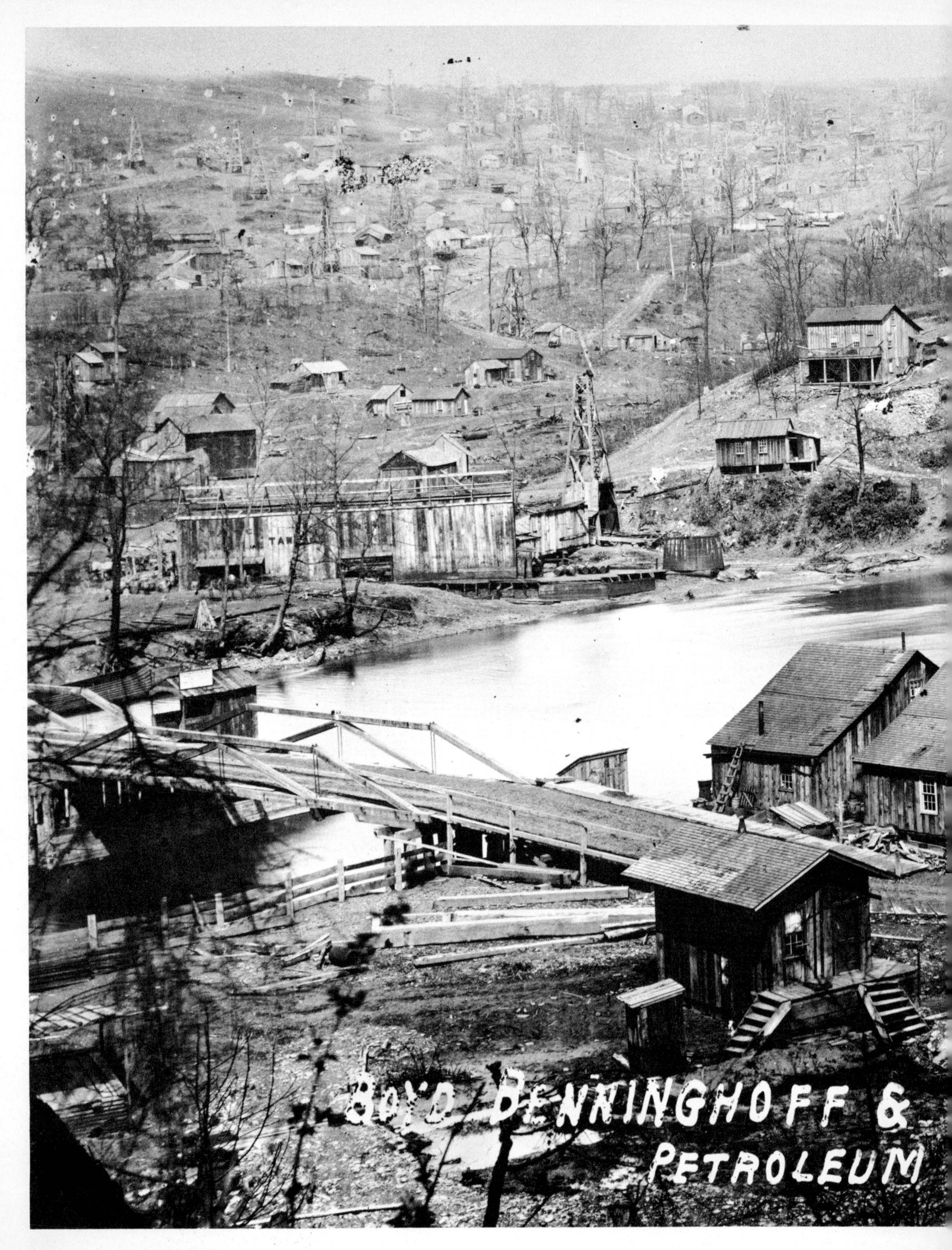
BOYD BENNINGHOFF &
PETROLEUM

OIL EXCHANGE HOTEL.
OIL EXCHANGE HOTEL.
LIVERY & FEED STABLE
VENSON FARMS
NTER 1864-8

Benninghoff Farm, Benninghoff Run

Wooden Storage Tanks, Benninghoff Run

For many years John Benninghoff had wrung a bare subsistence from the stony soil of his farm which lay across Benninghoff Run, Western Run, and Pioneer Run. As most of the farm was highland territory, operators did not look with favor upon it until the fall of 1865, when a completed well produced 300 barrels. Benninghoff was besieged by others who wanted leases. Small lots were taken at boom prices and one fourth the oil, and shortly a dozen wells were being drilled. Riches were thrust upon John Benninghoff, and soon he had an income of about $6,000 a day. His royalty in November 1866 exceeded $33,000. Distrustful of banks and bankers, Benninghoff purchased a safe in which he could keep his money at home. One night in January 1868, burglars entered the house and stole over $200,000, as described by the *Titusville Morning Herald* of January 17, 1868.

Horse-drawn Flatcar, Benninghoff Run

Robbery of John Benninghoff—$020,000 Stolen.

It will be seen by our special telegraph dispatches, that the house of John Benninghoff was felonously entered on Thursday night, about half past seven, and robbed of $200,000. Mr. Benninghoff resides in a farm house on the celebrated Benninghoff Run, about three-quarters of a mile from Oil Creek, west side, and a mile and a half distant from Petroleum Centre. His house stands near the road, not frequented much of late, either by travel or teams, and the nearest neighbor is a quarter of a mile distant. Mr. B. is a Pennsylvania German, a plain, hard-working farmer, and about sixty years of age, He has a large family of grown up children, many of them staying at home. The old gentleman all of a sudden, had riches thrust upon him by the discovery of oil on his sterile acres, and is one of our petroleum millionaires. We believe the first well was struck on his farm on Pioneer Run, in the fall of 1865; the production of his whole farm for last month was six hundred barrels a day, and ten new wells are now going down on it. Mr. B.'s income for December last, was reported at $40,000, but his sudden and dazzling fortune never made a fool of him, and they say he cares nothing for appearances, and wants to live as frugally and work as diligently as when he used to eat his bread in the sweat of his brow. But he had taken the notion into his head to be his own banker, and to deposit his bonds and greenbacks in his own safe and house. And now he has paid the penalty of very foolish and short-sighted temerity, and he may thank Heaven that he was not beaten or killed by the villains who robbed him. The whole community have known that Mr. Benninghoff kept his money—and a great deal of it—at home, and will not be surprised to learn of the perpetration of this villainy, much as they may pity his misfortune. The circumstances of the robbery as detailed in our dispatches, show careful premeditation and preparation, but with prompt and vigilant police effort, we believe the criminals can be traced or intercepted in their flight and a good deal of the booty recovered.

Completed in June 1866, this well flowed about 300 barrels a day. It belonged to Major S. M. Mills, the genial proprietor of the Moore House in Titusville.

King of the Hills Well, Stevenson Farm

James S. McCray Farm and the Egbert Tract, Petroleum Centre. Two of the most notable farms at Petroleum Centre—the richest producing area in 1866—were the McCray farm on the high bluff overlooking Oil Creek, and the Hyde & Egbert farm at its base. Companies and individuals strained to secure even the smallest lease.

In the spring of 1870 a 300 barrel well drilled near the McCray farm caused McCray to be besieged for leases. Derricks quickly appeared. Every well tapped the underlying pool. Within four months the daily production was 3,000 barrels, or an income for McCray of $9,000 a day. For this farm, McCray refused $1,000,000 in 1871.

Dr. A. G. Egbert, a young physician at Cherrytree village, bought the Davidson farm of 38 acres in 1860 for $2,600 and one half the oil. Because he could not pay for the farm, he sold a half interest to Charles Hyde, and it was called the Hyde & Egbert farm. The Jersey Well, completed in the spring of 1863, began flowing 350 barrels a day. Probably no parcel of ground of equal size yielded a larger return. At one time, there were over twenty-three flowing wells on the tract.

Columbia Oil Company Farm. The Story farm was purchased in 1859 by some Pittsburghers for $40,000, who leased lots to various operators and secured a number of good wells. In May 1861, they organized the Columbia Oil Company with a capital of $200,000. Andrew Carnegie was one of the heaviest stockholders. In July 1863, the first dividend of thirty per cent was declared, followed in August and September by two of twenty-five per cent and in October by one of fifty per cent. Four dividends, aggregating 160 per cent were declared at the end of the first six months of 1864. The capital of the company was now increased to $2,500,000.

The company was splendidly managed and had the best paying farm in the region. It built houses for the employees, erected machine shops, established a library, and maintained a band which became famous in the oil region.

Farm Manager's Office, Columbia Oil Company

Machine Shops, Columbia Oil Company

Columbia Cornet Band

TARR FARM, 1865

James Tarr's Residence, Tarr Farm

Boiler and Engine "For Sale Cheap," Tarr Farm

Group at Tarr Farm

Residence of John Blood

The Ferry at Blood Farm

South of the Story and Tarr farms, on both sides of the Creek, were John Blood's four hundred and forty acres. The farm was one of the earliest to be developed; in 1861 and 1862, when oil was almost valueless and ran to waste, it produced more oil than all of the farms in the region together. The sale of the Blood farm in April 1864, for $650,000 in greenbacks eclipsed all real estate transactions; it was the largest amount of cash that had ever been paid for any oil land up to that time. With three or four exceptions, it was the most productive farm in the entire oil region.

Twin Wells, Blood Farm

NIAGARA & PIERSON FARMS, CHERRYTREE RUN

Musicians and Oilmen, Hess Farm Wells, Cherrytree Run, 1864

Flowing into Oil Creek from th west, Cherrytree Run had more tha three hundred wells drilled along it course. Their success was not startlin but they were good producers. Prolifi wells on the Niagara & Pierson trac added materially to the wealth of th Phillips brothers—Isaac, Thomas W. and Samuel.

The Rynd farm of 300 acres, owne by John Rynd, was divided by O Creek and Cherrytree Run flowe through the western half into O Creek. A score of companies drille vigorously on Rynd farm. Five re fineries at one time operated on th farm.

Allemagoozelum Well, Cherrytree Run, 1869

Hess Farm Wells, Cherrytree Run, 1869

"Coal Oil Johnny's" Farm, 1864

John W. Steele

More familiarly known as "Coal Oil Johnny," John W. Steele was the adopted son of Culbertson McClintock and his wife. McClintock died before the oil rush. Mrs. McClintock leased a portion of the farm, excellent paying wells were struck, and she acquired considerable wealth. When she died in 1864, the twenty-year old Steele inherited the farm and cash estimated at $200,000. His daily income from oil was placed at $2,000 a day. Upon attaining his majority Steele decided to see the world and began a spending orgy that extended over the next twelve months. He went to Philadelphia and New York, became a victim of strong drink and bad company, and squandered his fortune. Because he derived his wealth from oil, newspapers nicknamed him "Coal Oil Johnny" and published the most sensational stories about his wild escapades.

An Oil Prince in Bankruptcy.—The Pitts burgh *Commercial* says that John W. Steele, famil iarly known as "Johnny" Steele, and somewha distinguished as an "oil prince," having for a considerable length of time enjoyed the princely income of $2,000 per day, on Thursday filed in the United States District Court, a voluntary petition in bankruptcy. Many of our reader will remember the romantic history of his ex ploits in the East, published some time ago during which he is reported to have squandered several hundred thousand dollars. After having "sowed his wild oats," and losing his oil farm he found himself in rather straightened circum stances, and was recently compelled to earn a living by driving an oil team. His indebtedness, as set forth in his petition, amounts to over $100,000 Some of the items are quite heavy, a few of which we note. To Henry W. Kanaga, of the Girar House, Philadelphia, he owes $19,824; to Wm A. Galbraith, Attorney-at-law, Erie, $10,000; J E. Caldwell & Co., Philadelphia, for jewelry $5,805; John D. Jones, for harness, $1,250; Wm Horn & Co., for cigars, $562; E. H. Conklin Philadelphia, liquors, $2,024; Phelan & Collen der, Philadelphia, for billiard tables, $1,500; to an unknown creditor, for oil paintings, $2,200 to the account for hats, $300. A considerabl amount of his indebtedness is for money borrow ed, notes, judgments, etc. When "Johnny" took a notion to rent a hotel for a few days h would do so; and whenever he saw anything tha pleased his fancy he was bound to have it regard less of cost. Perhaps no man in the United State ever squandered as much money in the sam space of time.

Bankruptcy

ON THE

DANGEROUS EXPLOSIBILITY

OF

PETROLIUM, COAL OILS,

CAMPHENE, SALT-PETRE, &c.

(From the Annual Report of the Smithsonian Institution for the year 1861-2.)

BY ZACHARIAH ALLEN.

PROVIDENCE:
SIDNEY S. RIDER, 17 WESTMINSTER STREET.
1862.

Ignorance of petroleum as an illuminant, the poor quality of refined oil, and the large number of deaths resulting from the use of coal oil all acted as a deterrent in expanding the market. Many warnings were published and distributed throughout the country.

The boom in drama on oil themes, set off by the exploits of "Coal Oil Johnny," continued into the 70's.

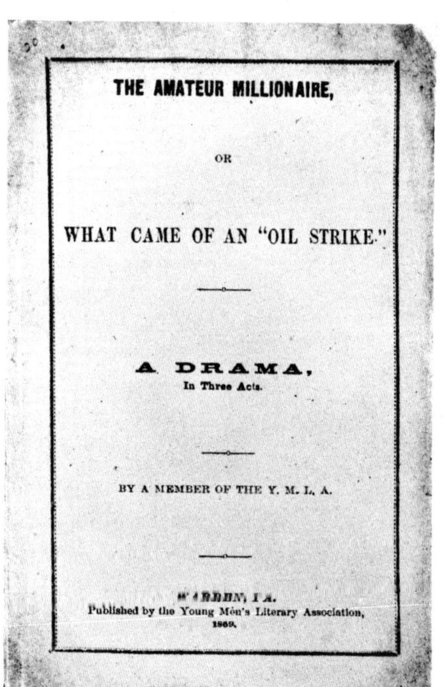

THE AMATEUR MILLIONAIRE,

OR

WHAT CAME OF AN "OIL STRIKE."

A DRAMA,
In Three Acts.

BY A MEMBER OF THE Y. M. L. A.

WARREN, PA.
Published by the Young Men's Literary Association,
1869.

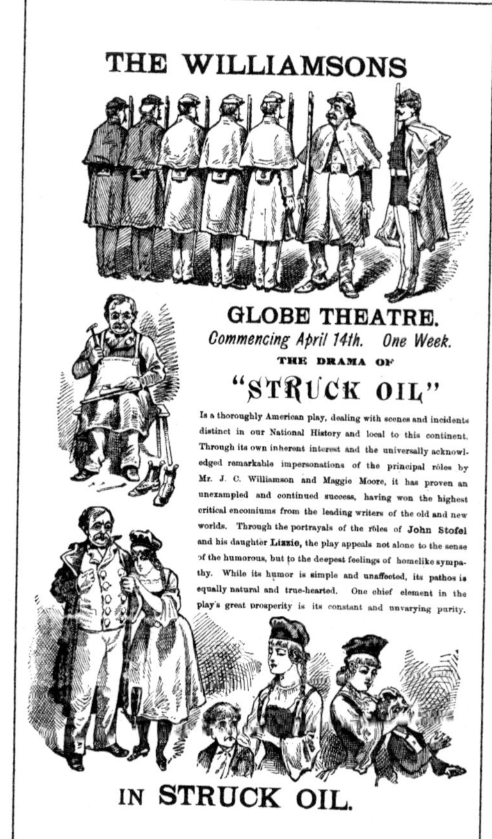

THE WILLIAMSONS

GLOBE THEATRE.
Commencing April 14th. One Week.
THE DRAMA OF
"STRUCK OIL"

Is a thoroughly American play, dealing with scenes and incidents distinct in our National History and local to this continent. Through its own inherent interest and the universally acknowledged remarkable impersonations of the principal rôles by Mr. J. C. Williamson and Maggie Moore, it has proven an unexampled and continued success, having won the highest critical encomiums from the leading writers of the old and new worlds. Through the portrayals of the rôles of **John Stofel** and his daughter **Lizzie**, the play appeals not alone to the sense of the humorous, but to the deepest feelings of homelike sympathy. While its humor is simple and unaffected, its pathos is equally natural and true-hearted. One chief element in the play's great prosperity is its constant and unvarying purity.

IN STRUCK OIL.

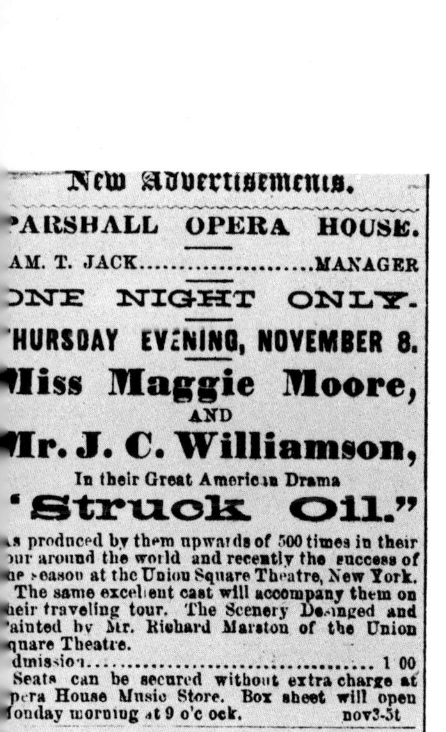

New Advertisements.

PARSHALL OPERA HOUSE.
SAM. T. JACK..........................MANAGER
ONE NIGHT ONLY.
THURSDAY EVENING, NOVEMBER 8.
Miss Maggie Moore,
AND
Mr. J. C. Williamson,
In their Great American Drama
"Struck Oil."
As produced by them upwards of 500 times in their tour around the world and recently the success of the season at the Union Square Theatre, New York.
The same excellent cast will accompany them on their traveling tour. The Scenery Designed and Painted by Mr. Richard Marston of the Union Square Theatre.
Admission.. 1 00
Seats can be secured without extra charge at Opera House Music Store. Box sheet will open Monday morning at 9 o'clock. nov3-5t

TO THE PUBLIC.

Not once, in its long career has an unfavorable comment been made upon its tone or language, but its moral composition has always been praised. The Williamsons, in their impersonations, offer their audiences truly lovely pictures of humble home-life and companionship between father and daughter, in Act I. In Act II. the stirring scenes of camp-life, with a humorous side and also a pathetic side, are depicted, while in Act III. the denouement is reached through scenes setting forth quickly reached prosperity and the result of steadfastness to the best impulses. The portrayals of **John Stofel**, by Mr. Williamson, and of **Lizzie Stofel**, by Maggie Moore, have won most glowing praise from all critics, and especially has the depiction of madness and gradual recovery of sanity by Mr. Williamson been most cordially and enthusiastically endorsed as a wonderful psychological study. With "Struck Oil" the Williamsons have made the grand tour of the world, performing in the drama over 1000 times in the principal cities of England, Ireland, Scotland, India, Australia and all the large cities in the United States, from Boston to San Francisco. Among their noted seasons in the piece may be reckoned over eight months in London, and four distinct engagements played at the Union Square Theatre and the Grand Opera House, New York, within fourteen months.

"Struck Oil" will be presented with the following admirable distribution of characters:

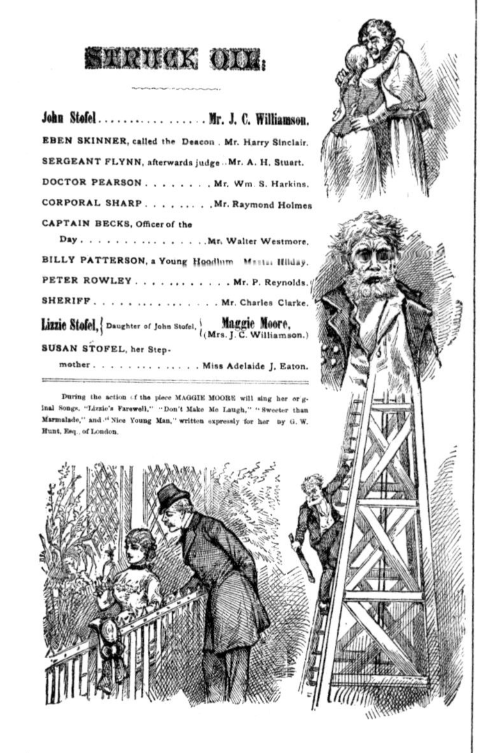

STRUCK OIL.

John Stofel .... **Mr. J. C. Williamson.**
EBEN SKINNER, called the Deacon . Mr. Harry Sinclair.
SERGEANT FLYNN, afterwards judge .. Mr. A. H. Stuart.
DOCTOR PEARSON . . . . Mr. Wm. S. Harkins.
CORPORAL SHARP . . . . Mr. Raymond Holmes
CAPTAIN BECKS, Officer of the Day . . . . Mr. Walter Westmore.
BILLY PATTERSON, a Young Hoodlum [illegible] Hilday.
PETER ROWLEY . . . . Mr. P. Reynolds.
SHERIFF . . . . Mr. Charles Clarke.
Lizzie Stofel, Daughter of John Stofel, **Maggie Moore,** (Mrs. J. C. Williamson.)
SUSAN STOFEL, her Step-mother . . . . Miss Adelaide J. Eaton.

During the action of the piece MAGGIE MOORE will sing her original Songs, "Lizzie's Farewell," "Don't Make Me Laugh," "Sweeter than Marmalade," and "Nice Young Man," written expressly for her by G. W. Hunt, Esq., of London.

Cherry Run

Cornplanter Run Wells

Flowing into Oil Creek from the west at the Clapp Farm, Cornplanter Run was a favorite place for drilling in 1864 and 1865

Little or no effort had been made prior to 1864 to extend drilling operations away from Oil Creek. With so many productive wells on Oil Creek and the supply exceeding the demand, operators did not need to seek any other place. During the summer of 1864, however, William Reed started drilling about two miles up Cherry Run, which flowed into Oil Creek from the east about three miles above Oil City. The valley had been almost entirely neglected. When down to the proper depth, Reed's well started flowing at the rate of 280 to 300 barrels a day. The Reed well opened up a new and untested area and precipitated a wild scramble for land along the entire length of Cherry Run. The rush was tremendous, the excitement great, and land once thought worthless sold for fabulous sums.

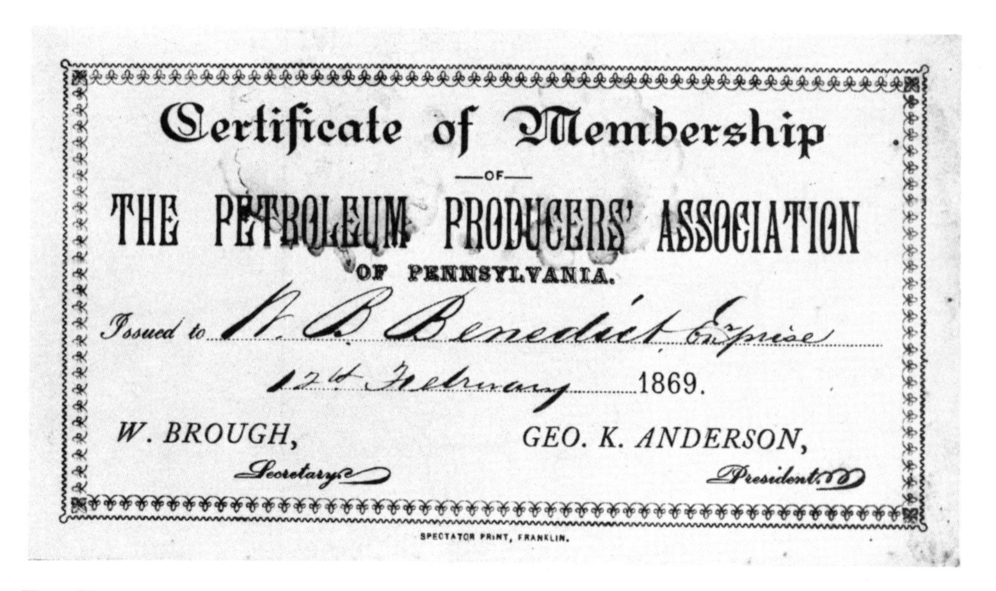

Certificate of Membership

—OF—

THE PETROLEUM PRODUCERS' ASSOCIATION

OF PENNSYLVANIA.

Issued to W. B. Benedict, Enterprise

12th February 1869.

W. BROUGH, Secretary

GEO. K. ANDERSON, President.

SPECTATOR PRINT, FRANKLIN.

The Petroleum Producers' Association. In order to oppose a bill in the state legislature to tax crude oil, to prepare monthly reports on production, and to attack the torpedo monopoly, the producers met in Oil City in February 1869 and formed the Petroleum Producers' Association—the first trade association of oilmen to be established.

CHAPTER III

PIONEER TOWNS ALONG OIL CREEK

EXCEPT for Titusville, which was a village of three or four hundred, there were no towns or cities along Oil Creek valley in 1859. In Titusville there were two churches and two taverns. "Pap" Hibbard kept the American Hotel where Drake stopped, the genial Peter Wilson ran the drugstore, and Brewer, Watson & Company not only conducted the general store but provided employment for many of the citizens at its lumber mills. There was no railroad, stages running twice a week to Erie brought the mail and connected this isolated community with the outside world. With the completion of the Drake well, Titusville became a whirlpool of excitement. It was the immediate destination of all who rushed to Oil Creek, and it quickly assumed a more prominent position than any other place in the region. By the summer of 1865 Titusville had developed into a bustling town of about 10,000 people.

Sixteen miles south, at the lower end of Oil Creek valley, there was in 1859 a little village called Cornplanter after the renowned chief of the Senecas. It consisted of a general store, a gristmill, several dwellings, and the Moran House. For years prior to Drake's discovery, Cornplanter had been a favorite overnight stopping place for hundreds of raftsmen, who ran lumber rafts to Pittsburgh from Coudersport, Warren, Tionesta, and other points up the Allegheny. With the striking of oil, Graff, Hasson & Company, which owned about one thousand acres of land on the east side of Cornplanter with an extensive front along the Allegheny, laid out lots for a town. Because of the rush, the village grew rapidly and in 1861 the name of the post office was changed from Cornplanter to Oil City. In 1862 it was made a borough, and the first newspaper, the *Oil City Register*, was established. By the spring of 1865 Oil City had become a prosperous and important shipping point with a population of about 8,000.

In the valley between Oil City and Titusville many oil towns sprang into existence as a result of the completion of some important well, or for some other reason, and then disappeared overnight without leaving much by which to be remembered other than the fact that each had been a "red-hot" oil town for a few weeks, months, or years.

PETROLIA;

OR,

THE OIL REGIONS

OF

THE UNITED STATES.

A Complete Descriptive and Guide Book,

CONTAINING

FULL INFORMATION

CONCERNING THE

GREAT PETROLEUM INTERESTS.

BY FRANC B. WILKIE.

Chicago:
JOHN R. WALSH & CO., Publishers,
1865.

GUIDE TO THE OIL REGIONS, 1865. As interest and excitement developed in all parts of the country over the events taking place along Oil Creek, many guide books were published for prospective investors and those who desired to visit the oil region.

Although not in the oil producing district, Corry was definitely an oil town. In 1859 there was no Corry; it was just a farm covered with trees, stumps, and two or three houses. Through it ran the Sunbury & Erie Railroad, later the Philadelphia & Erie and then the Pennsylvania, which afforded the only rail connection for shipping oil prior to May 1861. In that month the Atlantic & Great Western Railroad, which connected with the Erie at Salamanca for New York, extended its line in a southwesterly direction and crossed the tracks of the Philadelphia & Erie at Corry.

CORRY, PENNSYLVANIA, 1863

By October 1862, the Oil Creek Railroad was completed from Corry to Titusville. With three railroads, Corry grew like magic and became the most important transportation center in the oil region. Even though it had a population of 3,000 in 1865, it was known as "The City of Stumps."

"The Rush to the Cars at Corry"

Another reason for Corry's remarkable growth and prosperity was the fact that it had one of the largest and finest refineries in the oil region—the Downer Oil Works. Started in 1861 by Samuel Downer of Boston, the refinery when completed employed 150 men and had a refining capacity of approximately one million gallons a year.

GENERAL VIEW

North Side, Spring Street, Titusville, 1865

South Side, Spring Street, Titusville, 1865

TUSVILLE, 1864

Spring Street Looking West, Titusville, 1865

South Washington Street, Titusville, 1865

Hinkley Refinery, 1863

Opened in the spring of 1866, the Crittenden House on Pine Street was, for that day, the finest hotel in the oil region.

First Daily Newspaper in the Oil Region

Impressed by the rapid growth of Titusville and the development of the petroleum business, W. W. Bloss and H. C. Bloss came to Titusville in the spring of 1865 and established the Titusville *Morning Herald*, the first daily newspaper in the oil region. Its files furnish a most accurate and complete record of these early epoch-making events in the history of the petroleum industry.

Built in 1865, there was no larger nor more beautiful public hall in oildom than Corinthian Hall. It was here that P. T. Barnum, Josh Billings, Olive Logan, Mark Twain and many others lectured.

Corinthian Hall, Titusville

Brown's Band, Titusville, 1868

Mather's Boat

Mather's Wagon

Born at Heapbridge, England, in 1827, John A. Mather 'as one of a family of six sons and a daughter. His father erved as the superintendent of Ridgely & Sons paper mill, ne of the largest plants in England. In his youth Mather arted taking violin lessons and became an accomplished usician. He played for a number of English noblemen and 'as personally complimented by Queen Victoria.

When his father subsequently purchased a paper mill at lton Towers, Staffordshire, Mather thoroughly familiarized imself with the business, expecting to make it his life work. ut first, stimulated by letters from his two oldest brothers, 'ho had emigrated to the United States, he decided to visit em. In 1856, with his violin under his arm and $1000 in his ockets, he set out across the ocean.

Mather visited his brother Edmund who ran a summer otel in Pennsylvania at Sterret Gap. There he met an Engshman named Johnson, who traveled about the country king pictures. Mather was fascinated by photography and sked if he might go along and learn the business. Johnson ok him on and paid him five dollars a week and expenses. he pair visited many towns in West Virginia while Mather arned the art of making and finishing pictures. The newngled wet-plate process required considerable skill but lather soon became proficient.

Shortly thereafter Mather severed his connection with Johnson. He bought an outfit and struck out on his own through West Virginia, Maryland, and Ohio. He was working in Painesville in 1860 when he heard about Drake's famous oil well near Titusville, and the mad rush it had precipitated. Mather packed his equipment and arrived in Titusville in October 1860. After a week's stay he rented one half of a little building, put in a skylight and opened a gallery.

Mather spent the rest of his life in that town. For forty years he took pictures of pioneer oil men and their families, of famous oil wells, fields and farms, of teamsters, railroads, oil towns and countless aspects of Titusville and the surrounding oil regions. As the excitement moved down Oil Creek after 1860, Mather built a flat bottomed boat in which to carry his photographic equipment, for his pictures were made by the then revolutionary wet-plate method, which required immediate development of the negative. When activity moved away from Oil Creek, he used a horse and wagon. Wherever oil operators went, Mather was always close behind; he spared no labor in recording his exhaustive chronicle of the early petroleum industry.

MATHER'S OLD ESTABLISHED

hotographic Gallery

up one flight of stairs.

AMERICAN BLOCK, SPRING STREET, TITUSVILLE, PA.

LARGEST ROOMS IN TOWN.

Photographs, Porcelains, Ambrotypes, Gems, Doubl 'ositions, (superior) Card Pictures, and all other styles nown to the art. Copies from old pictures enlarged to ny size required. A large assortment of Albums, 'rames, Stereoscopic Cards, and large pictures of the Oil egions constantly on hand. Landscape views made t rder. Terms reasonable. oc20d1y

R. V. PIERCE, M. D.,

TITUSVILLE PA.

Consultation Rooms in Ralph's Brick Front Building,

lather's Advertisement, July 28, 1866

The Miller farm, about six miles be
low Titusville, lay on both sides of O
Creek; it was purchased in 1863 b
the Indian Rock Oil Company of Ne
York. It was excellent producing terr
tory. With the extension of the O
Creek Railroad down the valley i
1863, a flourishing town developed a
the terminus on Miller farm. It be
came a busy shipping point. Refinerie
were established, storage tanks erec
ed and, most important of all, a pip
line was built from Pithole to Mille
farm.

Indian Rock Oil Company Office, Miller Farm, 1865

Captain A. B. Funk bought in 1859 ıe McElhenny farms of 180 acres for ı,500 and one-fourth the oil, and ommenced drilling his first well in ebruary 1860. Early in May 1861, his ell, called "The Fountain," started owing 300 barrels a day. This was the rst well drilled to the third sand and ıe first on Oil Creek that really owed. Frederick Crocker drilled an-her well on the same farm about the ıme time and it flowed 1,000 barrels. he first great flowing well, the Em-ire, began flowing 3,000 barrels in ıe fall. These wells caused the terri-ory to boom. Derricks and engine-ouses studded the farm. Everybody anted to drill close to the spouters. town, Funkville, developed, flour-hed for a short time, and died.

McElheny Oil Company and Dalzell Petroleum Company Office, Petroleum Centre, 1863-1864

FUNKVILLE, 1867

Pioneer and Railroad Bridge, 1865

Located at the junction of Pionee Run and Oil Creek, Pioneer rapidl developed as an oil town. Feed stores offices, warehouses, hotels, storag tanks, and oil wells abounded.

Oilmen at the Erie Hotel in Pionee

Boughton Cooperage Shop, Pioneer, 1864

Brown's Hotel, Pioneer

McKinney Oil Office, Pioneer, 1864

Union and Hoskins Oil Company Offices, Pioneer, 1867

D. C. Clark's Store, Pioneer, 1864

Boarding House at Pioneer

Petroleum Centre, 1864

In 1863 the Central Petroleum Company of New York, organized by Frederic Prentice and George H. Bissell, leased the G. W. McClintock farm of 207 acres, eight miles from Titusville and eight miles from Oil City. The next year the company purchased the farm, drilled a number of wells, and granted leases for one-half the oil and a large bonus. Excellent wells were drilled. The company staked off a half-dozen streets and leased building lots at exorbitant prices. Since the town site was half way between Titusville and Oil City, they called it Petroleum Centre. Surrounded by some of the best oil-producing farms, Petroleum Centre was suddenly transformed into a lively town with 3,000 people, a

Central Petroleum Company Office, 1865

Refinery at Petroleum Centre, 1864

Central House, Petroleum Centre

ank, two churches, a theatre, a half-dozen hotels, a dozen ry goods stores, three or four livery stables, saloons, gamling dives, boarding houses, and scores of offices for brokers, shippers, and producers. With the influx of all sorts of people and an absence of government, the town soon eclipsed all others in wickedness.

J. J. Stoltz Boot & Shoe Shop, Petroleum Centre

George H. Bissell & Company Bank, Petroleum Centre

Washington Street, Petroleum Centr

Petroleum Centre Reunion, 1890

One of the favorite social diversions of the oil region was for oilmen and their families to assemble at some favorite spot for a picnic and reunion. In this particular case, thos from Petroleum Centre have gathered at a shady place alon the Allegheny River to enjoy a day of good fellowship.

McClintockville, 1861

McClintockville was located on the Hamilton McClintock farm a short distance above Oil City. It was the second farm to be leased for drilling purposes. For some years prior to drilling the Drake well, McClintock collected oil from a spring that bubbled up in the middle of the Creek, and around which he built a crib in order to prevent the oil from floating away on the current of the water. In the course of a year he collected several barrels and disposed of it with some profit to the surrounding farmers. The third well to produce oil on the Creek was drilled in this spring. The village, which took its name from the farm, had at one time a hotel, several boarding houses, and two or three refineries.

Street Scene, Tarr Farm

Street Scene, Rouseville, about 1867

Rouseville, 1867

Situated on the Archbald Buchanan farm at the mouth of Cherry Run and about three miles above Oil City, Rouseville was named after Henry Rouse of Warren, Pennsylvania, who died in 1861 as a result of burns received in the first great oil well fire in the region. Rouseville grew swiftly and for a time was the headquarters of the oil industry. Churches and schools were established; houses were built up Cherry Run; wells and tanks covered the flats, and stores and shops multiplied. Ida M. Tarbell lived here as a young girl. Rouseville reached its height in the early seventies.

Main Street, Oil City, 1861

OIL CITY

Because of its advantageous locatio at the junction of Oil Creek and th Allegheny River, many oil men mad their headquarters at Oil City and i rapidly developed as a shipping point In the fall of 1863 the Michigan Roc Oil Company, which owned the lan on the west side of Oil Creek at Oi City, divided its extensive river fron into lots suitable for oil landings an storage yards. Different firms bough them and soon warehouses, offices supply stores, and tool and coope shops crowded the river front. By Oc tober there were at least twenty oi landings and the capital investe amounted to about $10,000,000.

In the spring of 1863 Cottage Hil east of Oil Creek, was laid out and i was not long until a number of house were erected. A mania for buildin prevailed, and the sound of hamme and saw could be heard everywher from morning until night. As in othe towns, the buildings were hurriedl constructed without any thought c architectural beauty or permanence.

Long Bridge over the Allegheny River, Oil City, 1864

Hogback Hill, Oil City, 1863-1864

OIL - CITY

DAILY DERRICK.

VOL. 1. OIL CITY, PA., MONDAY, SEPT. 11, 1871. NO. 1.

PETROLEUM HOUSE.
Free Omnibus to and from all Trains.
BILLIARD AND BATH ROOMS CONNECTED WITH HOTEL.
C. W STAATS, Prop.

JAMESTOWN HOUSE,
A. M. SHERMAN, Proprietor.
Corner Main and Second Sts.,
JAMESTOWN, - - - NEW YORK.
Barber Shop and Bath Rooms. Large Parlors and Fine Sample Rooms. General Stage Office. TRY ME. sep11tf

UNION OUSE.
GEORGE STUBLER, Proprietor.
Elm Street,
OIL CITY, - - PENN.
sep11tf

M'CRAY H USE.
CHAS M. BLACKAN, Prop'r.
Washington Street,
PETROLEUM CENTRE, - - PENN.
This house is centrally located, and the General Headquarters of oil men. sep11tf

STEEL HOUSE,
Seneca Street, Oil City, Pa.
A FIRST-CLASS HOUSE,
Rooms and Beds Newly Furnished.
Table cannot be excelled.
J. M. STEEL, Proprietor

GIFFORD HOUSE,
H. C. FELTON, Proprietor.
JAMESTOWN, - - - NEW YORK.
Largest and best managed house in the [illegible]
T. CLUNEY, Manager. sep11tf

OCCIDENTAL HOTEL,
HARRY CHAMBERLIN, Prop'r.
MEADVILLE, - - - - PENN
Newly REFITTED AND REFURNISHED [illegible] for Commercial Travelers. Large Sample Rooms. sep11tf

EAGLE HOUSE,
S. T. JACK, Proprietor.
ELM STREET, OIL CITY, Pa.
The accommodations are second to none in the city.
TERMS—Transient boarders $1,50 per day. Table boarders $5 per week.

ABBOTT HUOSE,
FORMERLY
CRITT NDEN HOUSE,
Newly [illegible]ted and Re-Furnished
SAMPLE ROOMS.
MATHEWS, Proprietor,
TITUSVILLE, Pa.

HOUSE,
[illegible]LEUM CENTRE, PA.

DUNCAN HOUSE,
[illegible] CITY, PA.
[illegible] and Allegheny River Rail
CROSS & McOMBER, Prop'rs.

CITY EXCHANGE HOTEL.

BY TELEGRAPH.

NEW YORK.

Reception of Grand Duke Alexis—Killed—Annual Fair—Preparations for the Execution of Buckhart.

NEW YORK, Sept. 9.

Arrangements for the reception of the Grand Duke Alexis, which takes place early in October, are now taking a definite shape and attracting much attention. It is stated that Admiral Porter will command the fleet destined to receive the distinguished visitor, and that Admiral Stringham and suit of the Navy Yard will probably participate.

Jas. Dunn, laborer, aged twenty-five years, while at work raising the roof of No. 426 West street, yesterday afternoon, was instantly killed by a large timber falling on his head.

The twenty-seventh annual fair of the Society of Agriculturel and Horticulture of Westchester county commences near White Plains on Tuesday next.

The society offers $6,500 in premiums for the fastest horse. $600 in books, medals, silver ware and money will also be presented to the successful competitors in various departments.

WHITE PLAIN, N. Y., Sep. 9.

A fence has been constructed between the court house and jail here to enclose the scaffold on which Buckhart is condemned to be hung next Friday. A force of 200 military and a large number of special police have been selected for the occasion.

KENTUCKY.

Investigation of Railroad Accident—Murderer Pardoned—Coming Races.

LOUISVILLE, Sept. 9.

Coroner Webb has given notice that he will thoroughly investigate the recent accident on the L. & N. R. R., and if found there was any neglect and malfeasance he will cause the arrest of the guilty party.

John H. Morton was in jail for several months on conviction of killing Powers in a bawdy house was discharged to-day on pardon from Gov. Leslie.

The Judge of the Circuit Court decided this morning that Carl Grosse, the wife murderer was entitled to bail in $10,000.

LEXINGTON, KY., Sept. 9.

Pool selling for the coming races, commencing on Monday, took place at Broadway hotel this afternoon and resulted in but small sales. Jackson's Ballnza being the favorite for the one-mile dash, eight horses, sold for $50 against the fild. Others sold at about the same average. Bids were not spirited and the sales were far below the general expectation. Longfellow will run in the two-mile race next Tuesday.

CHICAGO, 9, 1871.

The Chicago Republican vigorously denis that it is to be bought by the democracy and made a party organ, though it is generally believed that negotiations for that purpose were stopped by the lack of necessary, money. It is now reported that the democracy intend purchasing the Times, which it it also claimed represented the party very faithfully, and there is a report that a company with 290,000 dollrs stock has been formed to start a new democratic paper.

OHIO.

McGehan Trial—Important Evidence.

DAYTON, Sept. 9.

PETROLEUM MARKETS.

Sales and Re-Sales of Crude Petroleum Delivered Upon Cars.

Office of the OIL CITY DAILY DERRICK, Saturday, Sep. 9, 1871.

The market to-day has been exceedingly quiet, but steady for Oil City delivery. 4 65 was paid for several lots of 47 gravity, but later in the day buyers at these prices were scarce, and 4 59 was the best offer. There is nothing doing in Oil City futures.

On the lower road there was some little inquiring for future sales of 4 50 being made at 4 60 and sellers thirty at 4 60. 4 65 was also bid for spot or regular. For round lots 4 70 was generally asked and no sales were reported on upper road. Nothing of importance was done, 4 35 being bid for spot or regular, while 4 37½ was asked.

There was more disposition to sell than to buy, however. Refined oil is weak and declining and there is every prospect of a lower market next week.

We quote.

1000 Bbls.	47 gravity	Oil City Pa........	4 80
3000 "	40 to 47 "	spot "	4 22½
1000 "	40 to 47 "	" "	4 35
1000 "	47 gravity at Oil City........		4 75
2000 "	47 "	" "	4 90
1000 "	47 "	spot "	4 85
1000 "	47 "	" "	4 85
3000 "	sellers 60 days "	......	4 80
800 "	47 gravity	......	4 70
5000	Lower road S. O. 60 days..		4 10
5000 "	" "	60........	4 60
5500 "	" "	60........	4 60
2000 "	"	S. O. 30 days........	4 65
1000 "	"	" 30 "	4 65
	8 cars 47 gravity Rouseville..		4 65
1000 "	spot on Upper road	....	4 37½
500 "	"	"	4 35
3000 "	P. T.	"	4 22½

Refined B. C. Oil, [illegible]

Titusville Market.

TITUSVILLE, Sept. 9—5 p. m.

The market has been quiet and dull to-day. Not much disposition to operate. We hear of one [illegible] 5000 barrels regular at $4 25.

Rouseville Market.

ROUSEVILLE, Sept 9—5 p. m.

The market was very quiet and dull to-day; $4 60@4 65 was offered and $4 70 bid.

Tidiou Market.

TIDIOUTE, Sept. 9—5 p. m.

The market was dull to-day. Sales at $4 45 @4 50.

Petroleum Centre Market.

PETROLEUM CENTRE, Sept. 9—5 p. m.

The market was quiet to-day. Sales at $4 60@4 65.

STOCK MARKDT.

MONEY AND STOCK.

NEW YORK, Sept. 9.

The events of the week, in financial circles, were important in some respects. Stock Exchange opened with a vigor, and a further advance was made in leading speculative shares. Pacific Mail has been very active, and at one time touched 58 58. The dealings in Toledo & Wabash shares formed a prominent feature of the market. On several days of the week, at the highest excitement in stocks, the price rose to 68 5-8. New York Central advanced to 103 3-8. Lake Shore touched 114 1-2. Amid the excitement Northwestern and St. Paul shares made a sharp advance on increased business. At the close of the week the market became heavy, and there was a strong decline of prices on the general list.

In the Government Bond market, prices

CHICAGO.

The Cheney Difficulty.

CHICAGO, Sept 10.

An immense audience gathered at Christ church to see the result of the Bishop's promised visit. The rector, Mr. Cheney, officiated as usual and preached a powerful sermon from the text, "I must work the work of Him that sent me." The sermon contained no direct reference to the presence of the Bishop or the troubles of the church, and at the close the congregation was dismissed much disappointed, as it was expected that the Bishop would appear. It seems that he came according to appointment and was met by the vestry, who informed him that he could not officiate unless Mr. Cheney also officiated, whereupon the Bishop left. Many are displeased with Cheney for not yielding so much as to take a seat in the congregation and allowing the Bishop to confirm the candidate. The Bishop has now exhausted every means and it is thought he will probably bring a civil suit for the church property.

CALIFORNIA.

Serenade of Governor-elect Booth—Fire.

SAN FRANCISCO, Sep. 9.

Hon. Newton Booth, Governor elect, arrived in this city yesterday and was serenaded at the Grand Hotel last evening.

The extensive barrel and stove factory of Flint, Peabody & Co. was destroyed by fire last night. The loss will be $150,000; fully insured.

Cable News.

GERMANY.

Alliance Between Austria and Prussia.

BERLIN, Sep 10.

The Prussian Cross Gazette officially declares the result of the recent conference at Gastein between the Emperors of Germany and Austria to have been the conclusion of an alliance offensive and defensive between Austria and Germany, on the following basis: 1st, the maintainance intact of the Austrian empire in its present condition. 2d, the maintainance of the present empire of Germany; and, third, the combination of both powers for a joint opposition against aggression from any other power whatever.

FRANCE.

Evacuation of Prussians—Alarming Conspiracy—Death of Admiral Bonet.

PARIS, Sept. 10.

The evacuation of the departments near Paris was commenced to-day by the Prussian troops. The occasion was celebrated by a dinner given by President Thiers at the Executive House. General Von Manteufel, Gen. Chenzy and Gen Ducrot were present along with a number of well known leaders in diplomatic circles, and warm personal friends of President Thiers. The occasion was characterized by much harmony and cordiality.

TOULON, Sept. 10.

A conspiracy of most alarming proportions and character, has been discovered by the authorities of this city. The discovery, fortunately, occurred in time to prevent the consummation of what bid fair to be one of the most deeply-laid and villainous plots which has ever come to the notice of the authorities. A plan had been concocted by a number of persons in this city, supposed to be in some manner

WIN OR BROS.

DEALERS IN

GENEPAL HARDWARE

And Oil Well Supplies,

TUBING, CASING, ROPE.

OAKUM, BELLOWS,

ANVILS &c. &c.

THE LARGEST STOCK

—OF—

Builder's Hardware

IN THE

OIL REGIONS.

Jewelry & Fancy Articles,
GOLD WARE.
Double Plated Sets, Horn & Melton Gold,
All new Styles and Patterns, just opening; also a great variety of Bracelets, Neck-Laces, Chains and Opera Chains, at popular prices.
SAM. M. RIESER,
sep11tf 37 Centre Street.

BLACKSMITHING.
DAVID L. TRAX,
General Blacksmith, Machinist,
AND MANUFACTURER OF
OIL WELL TOOLS,
SENECA STREET, OIL CITY, PA.
Having recently made important additions to my establishment, of the most approved machinery for the manufacture of
OIL WELL TOOLS,
I am now ready to give the public the full benefit of the same. [illegible]
STEAM HAMMER,
for Forging and Welding Oil Well Tools and other work in that line.
[illegible] OIL WELL TOOLS [illegible]
HORSE SHOEING AND REPAIRING.
WAGONS IRONED.
And all work in my line executed to the best style and on short notice.
Sep 9-11 D. L. TRAX.

W. F. FOX & CO.,
LUMBER YARD
AN
PLANING MILL.
Sash Doors and Blinds.
OFFICE AND YARD:
North Seneca Street.

SLIGO IRON STORE.
I. B. JACOBS,
DEALER IN
SLIGO, NORWAY,
AND
COMMON IRON,
English and American Steel,
ANVILS, NAILS, DRILLING ROPE.
Stoves and Tinware,
And General Housekeeping Articles.
Terra Cotta Ware,
AND
SEWER PIPE FOR SALE.
9 Centre St., Oil City, Pa.
sep11tf

"Specialties Prepared by"
H. WILLIAMS & CO.
Factors and Dealers
IN
Pharmaceutical Products,
NO. 7 CENTRE ST.,
OIL CITY, PA.
SACHET POWDERS,
MILLE FLEUR, VIOLET,
HELIOTROPE, FRANGIPANI,
QUININE OF GLYCERINE
HAIR TONIC,
Possessing the tonic properties of Quinine with the softening properties of Glycerine, rendering the Hair soft and beautiful, promoting its healthy growth, and preventing it from falling out.
Cologne and Extracts.
FLAVORING EXTRACTS,
VANILLA, LEMON,
FOR FLAVORING
JELLIES, ICE CREAM,
CAKES, BLANC MANGE,
PUDDINGS, SAUCES, PASTRY, ETC., ETC.
Prepared with the greatest care from the best materials, this Extract is unsurpassed in strength, Purity, and Delicacy of Flavor.
PLATE POWDER FOR
CLEANING SILVER.
CAMPHORATED DENTRIFICE.
An invaluable article for Preserving

A portion of the first page of the first issue of the *Oil City Derrick*, which appeared on September 11, 1871, with Coleman E. Bishop as the editor. In 1885 Patrick C. Boyle, former oil field laborer, teamster, oil scout, and newspaper reporter, bought the *Derrick* and made it "The Organ of Oil." The *Derrick* covered all the important developments in the oil industry, and is an indispensable source of information on the history of the industry.

Union Station, Oil City. Oil City had no railroad connection with the outside world until March 1865, when the Atlantic & Great Western Railroad completed a branch line from Meadville via Franklin to Oil City. In the fall of 1866 the Farmers Railroad connected Oil City with Petroleum Centre, and gave the "Hub of Oildom," as Oil City was called, an outlet over the Oil Creek Railroad.

Agitator at Brundred Refinery, Oil City

Commercial Hotel, Oil City

T. M. George's Keystone House, Franklin. The old India trading post of Franklin, at the junction of French Cree and the Allegheny, became important as an oil town.

Teamster on Plank Road

Teamsters Fording Oil Creek

CHAPTER IV

TRANSPORTING OIL FROM OIL CREEK

THE facilities for shipping oil from Oil Creek prior to 1862 were crude and inadequate except when the water in Oil Creek was high enough to permit flatboating. High water stages averaged less than six months a year, and during the rest of the time oil had to be hauled to the nearest railroad stations—Corry, Union Mills, and Garland—about twenty to twenty-five miles north of Titusville. Thousands of teams, therefore, were regularly engaged in hauling oil to these shipping points.

An alternative to hauling oil was the pond freshet, which lumbermen had used for years to raft their logs down Oil Creek when the water was too low to permit navigation. To create an artificial freshet there were at least seventeen sawmills with dams on the principal branches of Oil Creek, some of which were as much as ten miles above Titusville. Through a system of floodgates, the water could be held until a sufficient quantity had backed up; then it was let loose, thereby making a stage of water below sufficient to float logs down Oil Creek to the Allegheny. Since this method was cheaper and quicker than teaming, the oilmen simply appropriated the idea of the pond freshet. After all boats along the Creek had been loaded with oil and a toll collected on each barrel, word was sent up the Creek to cut the dams. When all the dams had been cut, the water would rush down Oil Creek, causing a rise of twenty to thirty inches. At the proper moment the boatmen would cut loose their lines and, within a short time, 150 to 200 flatboats, large and small, loaded with ten, twenty, or thirty thousand barrels of oil, either in bulk or barrels, would be floating along endways and sideways on a rushing flood and wildly fighting their way down Oil Creek to Oil City. The pond freshet always involved a heavy loss of oil; a third of it was lost by leakage before the boats started, and another third was lost before reaching Pittsburgh.

In 1862 the Oil Creek Railroad was built from Corry to Titusville, a distance of about twenty-seven miles; within the next two years the road was extended down Oil Creek to Shaffer farm and later to Petroleum Centre. The completion of the railroad lessened the necessity for hauling oil long distances over frightful roads and led to the abandonment of the pond freshet.

Bad Roads, Sherman Well, Oil Creek, 1863

Loading Oil at Funkville

Filling Barrels With Oil

Filling Flatboats with Oil

One of the most destructive pond :eshets occurred on May 31, 1864, hen one of the boats swung cross- ise of the bridge pier at Oil City. ncoming boats smashed into it, caus- g a jam and thousands of dollars orth of damage.

As early as April 1860, the steamer *enango* carried the first load of etroleum to Pittsburgh and within vo years there were fifteen steam- oats and towboats plying between Oil ity and Pittsburgh, each having an verage capacity of about 800 barrels. he steamboats averaged about three ips a week when the river was in ood navigating condition and the wboats two. In 1862 Jacob Jay Van- ergrift of Pittsburgh started the bulk oat business; his boats were about ghty feet long, fourteen feet wide, d three feet deep, each with a apacity of about 400 barrels.

Oil barges anchored in the Allegheny iver at Oil City, some of which are reparing to go to Pittsburgh with oil; thers have come up the river with npty barrels and supplies for the oil egion.

Spectators Watch Pond Freshet, Oil City, May 1864

The Oil Fleet at Oil City

Transporting Barrels on Oil Creek

Getting Flatboats Up Stream. When the water in Oil Creek was high enough to permit flatboating, which was about six months of the year, the Creek was extensively used for transporting oil to market.

Packet Express Boat
Vithout any roads or means of transportation, visitors to the oil region rode up and down Oil Creek in "packet express" boats as the one above was called.

Passenger Packet Going Down Oil Creek

Teaming on Oil Creek

MAP OF THE OIL REGION, showing railroad lines entering the Oil Creek region in 1865.

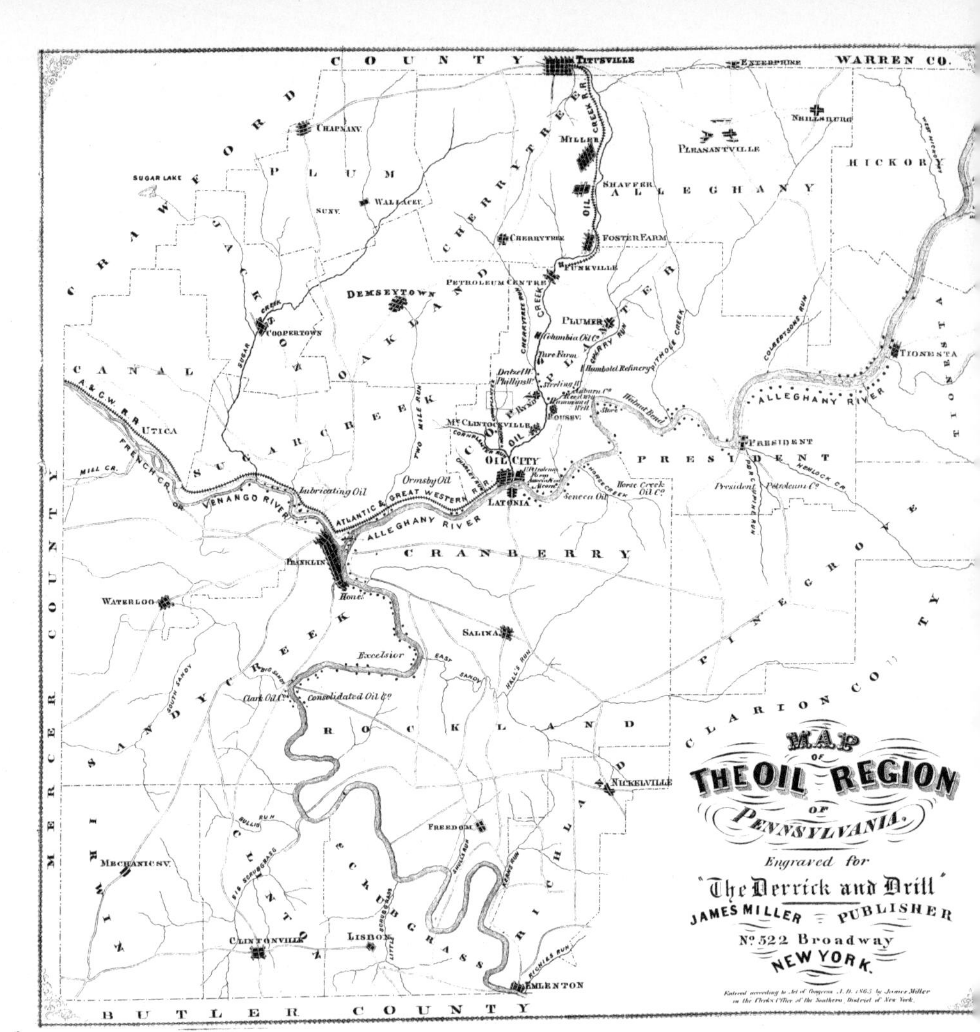

Built from Corry to Titusville in 1862, the Oil Creek Railroad was extended down Oil Creek to Petroleum Centre in 1866. From the beginning, it had an overwhelming volume of business, and it proved to be one of the best paying lines in the country. After deducting all expenditures for the first fourteen months, the directors declared a dividend of 25 per cent. For 1864, it paid a net profit of 53 per cent.

Locomotive "Oil City," Oil Creek Railroad

The Locomotives of the Oil Creek Railroad, at Rouseville, Petroleum Centre, and the Pioneer bridge.

The first oil cars were regular flat- ırs on which oil barrels stood on end. he picture above is of the Oil Creek ailroad near Rouseville.

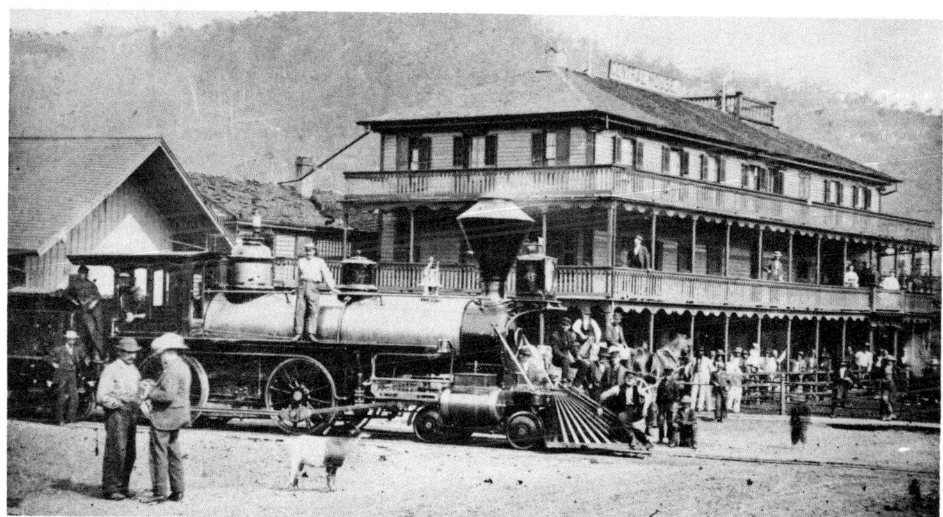

Oil Creek Railroad Locomotive at Petroleum Centre

Oil Creek Railroad Locomotive on the Bridge, Pioneer

CHAPTER V

PITHOLE

LITTLE or no effort had been made prior to 1864 to extend drilling operations away from Oil Creek. Beginning in 1864, however, with the expansion of the market, increased confidence in the oil business, a marked improvement in the price of oil and the opportunity for enormous profit, operators began "wildcatting" in areas away from the Creek.

In the spring of 1864 the United States Petroleum Company leased the farm of Thomas Holmden on Pithole Creek and commenced to drill in a wild and secluded spot. On January 7, 1865, the United States or Frazier well startled the oil region when it began flowing 250 barrels a day. A general stampede to Pithole Creek followed. People came by the hundreds and thousands. As other wells were successfully completed, the excitement mounted until the climax was reached in September, when the daily production amounted to about 6,000, or two-thirds of all the oil produced.

In May, a town was laid out in the wilderness on th Holmden farm, and it came to be known as Pithole. B September the town had two banks, two telegraph offices, daily newspaper, a waterworks system, two churches, theatre, over fifty hotels, the third largest post office in Pen sylvania, and about 15,000 inhabitants. Pithole was oi dom's greatest boom town.

The rise of Pithole was swift and amazing but its declin was even more breath-taking. When some of the big well stopped flowing in August and an increasing number of dr holes were drilled, production fell off sharply. Operator speculators, businessmen and others quickly departed. B January 1866, Pithole was a deserted village. A series c disastrous fires razed most of the buildings; others wer torn down and moved to other towns. The derricks wer soon cleared away and today silence reigns over the grass land which once provided oildom's greatest excitement.

UNITED STATES OR FRAZIER WELL. The completion of this wildcat well in January 1865, started the stampede to Pithole Cree

Astor House, the first hotel at Pithole
—built in about a day.

JOHN WILKES BOOTH

During the summer and fall of
864 John Wilkes Booth visited Mead-
ille, Franklin, Pithole Creek and other
laces in the oil region. He owned a
hirteenth interest in the Homestead
well, for which he paid $15,000, but
old out within a short time and left
he region.

United States Petroleum Company Office

HOMESTEAD WELL

Completed in April 1865, the Homestead well on the Hyner farm was the second to be drilled at Pithole. It pumped fifty barrels a day, later five hundred.

Pithole and Balltown

ONE SHARE IN THE
30 SHARES. 16 WELLS.
ELLSWORTH OIL WELL ASSOCIATION
OF PITHOLE CREEK.

This Certifies, that Horace E. Wells, his heirs or assigns, is entitled to the One-Thirtieth Interest in Twenty Thirty-Seconds, in Sixteen Wells, on the HOLMDEN, BALL, MC KINNEY, AND MOREY FARMS, Pithole Creek, Venango County, Penna., purchased by the Ellsworth Oil Well Association, and described in the Articles of Association dated November Fourteenth, 1865. Transferable only by written notice to the Trustees.

No.

Observer Presses, Erie, Pa.

Trustees.

The stock certificate illustrates the speculative craze which took place during the summer and fall of 1865 and how small the shares were split.

Other wells were drilled and completed in rapid succession at Pithole. The Grant well, named in honor of the hero of Appomattox, started flowing in August at 200 barrels and soon spouted 1,200.

Grant Well Office

First Street, Pithole

Within three months' time, between May and September 1865, Pithole had come into existence with a population of about 15,000.

Looking over the roof of Duncan & Company's office, one can see the Bonta House.

Eureka Well

Typical Wildcat at Pithole

Charles Pratt
Captain of the Pithole
Swordsman's Club

Organized for social purposes in 1865, the Swordsman's Club quickly attained a wide reputation. The members elegantly furnished some rooms, and good fellowship prevailed at all their meetings. The motto of the club was "R.C.T."—meaning rum, cards, and tobacco. The club gave elaborate concerts and balls to which all the best people of oildom were invited; they were the outstanding social events at Pithole.

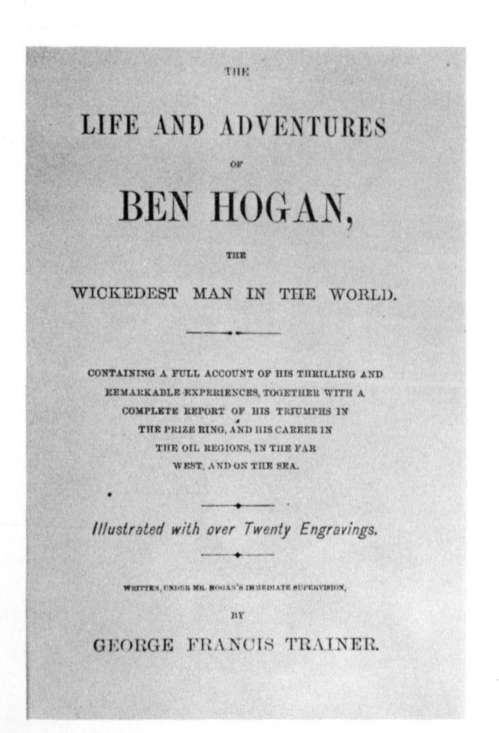

THE

LIFE AND ADVENTURES

OF

BEN HOGAN,

THE

WICKEDEST MAN IN THE WORLD.

CONTAINING A FULL ACCOUNT OF HIS THRILLING AND REMARKABLE EXPERIENCES, TOGETHER WITH A COMPLETE REPORT OF HIS TRIUMPHS IN THE PRIZE RING, AND HIS CAREER IN THE OIL REGIONS, IN THE FAR WEST, AND ON THE SEA.

Illustrated with over Twenty Engravings.

WRITTEN, UNDER MR. HOGAN'S IMMEDIATE SUPERVISION,

BY

GEORGE FRANCIS TRAINER.

Bonta House, erected at a cost of $80,000, was one of the most elegant hotels a Pithole. The other famous hotel, the Danforth House, was built on a lot leased fo $14,000. The building later sold for $16.

Ben Hogan
"The Wickedest Man in the World"

Although born in Switzerland, Ben Hogan spent the early part of his life in Germany. When eleven years of age, his parents brought him to the United States. During the Civil War, Hogan and some of his associates bought a large vessel with which to run the blockade at Charleston. Before the Union Navy forced them out of business, Hogan reported that they made $75,000 in gold and silver. Hogan then turned to enlisting soldiers for the Union Army, and claimed he made as much as $10,000 a month.

"French Kate" was associated wit Ben Hogan in his business enterprise in the oil region.

With this kind of background, Hoga arrived at Pithole in 1865 and opene a palatial sporting-house, the receipt from which often totaled $1,000 day. Hogan followed the tide of oi developments and opened a "free an easy" at Tidioute, Titusville, Parker Landing, Petrolia, Bullion, and Brad ford. He was the most notorious char acter in the oil region.

Pithole Pioneers Reunion. In December 1892, oilmen of Pithole's palmy days gathered at the New Brunswick Hotel in Titusville for a banquet and reunion.

Dining Hall at Pithole

Holmden Street and Danforth House, 1865

1895 Corner of Holmden and First Streets, Pithole 1934

CHAPTER VI

REVOLUTIONIZING THE PRODUCTION AND TRANSPORTATION OF OIL

THE years 1865 and 1866 witnessed a complete change in the mode of transporting oil to market. Instead of the once familiar piles of barrels at the wells and shipping points and long processions of teams and wagons, the oil region now had miles of pipe lines, tank cars along the sidetracks, and fast freight service to the seaboard.

The extension of the railroads into the oil region made possible the introduction of the tank car, which Amos Densmore of Miller Farm devised during the summer of 1865. On an ordinary flatcar he built two wooden tanks, one on each end over the trucks. With each tank car containing forty to forty-five barrels of oil, Densmore sent the car to New York. The experiment proved successful, other tank cars were built and by the spring of 1866 hundreds of them were in use. This type prevailed until the iron boiler tank car first appeared in February 1869.

Several early experiments had demonstrated the practicability of transporting oil through pipe lines, but their successful operation over greater distances did not come until the summer of 1865. The unsatisfactory condition of the roads, the exorbitant charges of the teamsters, the waste, and the production of oil faster than it could be hauled away from Pithole influenced Samuel Van Syckel, an oil buyer, to lay a two-inch pipe from Pithole to Miller Farm on the Oil Creek Railroad about five miles away. Completed in October 1865, Van Syckel's pipe line pumped about eighty barrels of oil through the pipe in an hour. The experiment worked perfectly and, within a short time, many other pipe lines were in operation.

Another innovation, which revolutionized the production of oil, was the invention of the torpedo by Colonel E. A. L. Roberts. In January 1865, Roberts made the first public demonstration with his torpedo in the Ladies' well on Watson Flats below Titusville. Owing to his remarkable success in increasing the production of oil, the demand for Roberts' torpedo was enormous, thousands were sold, and his method became one of the established practices in oil production.

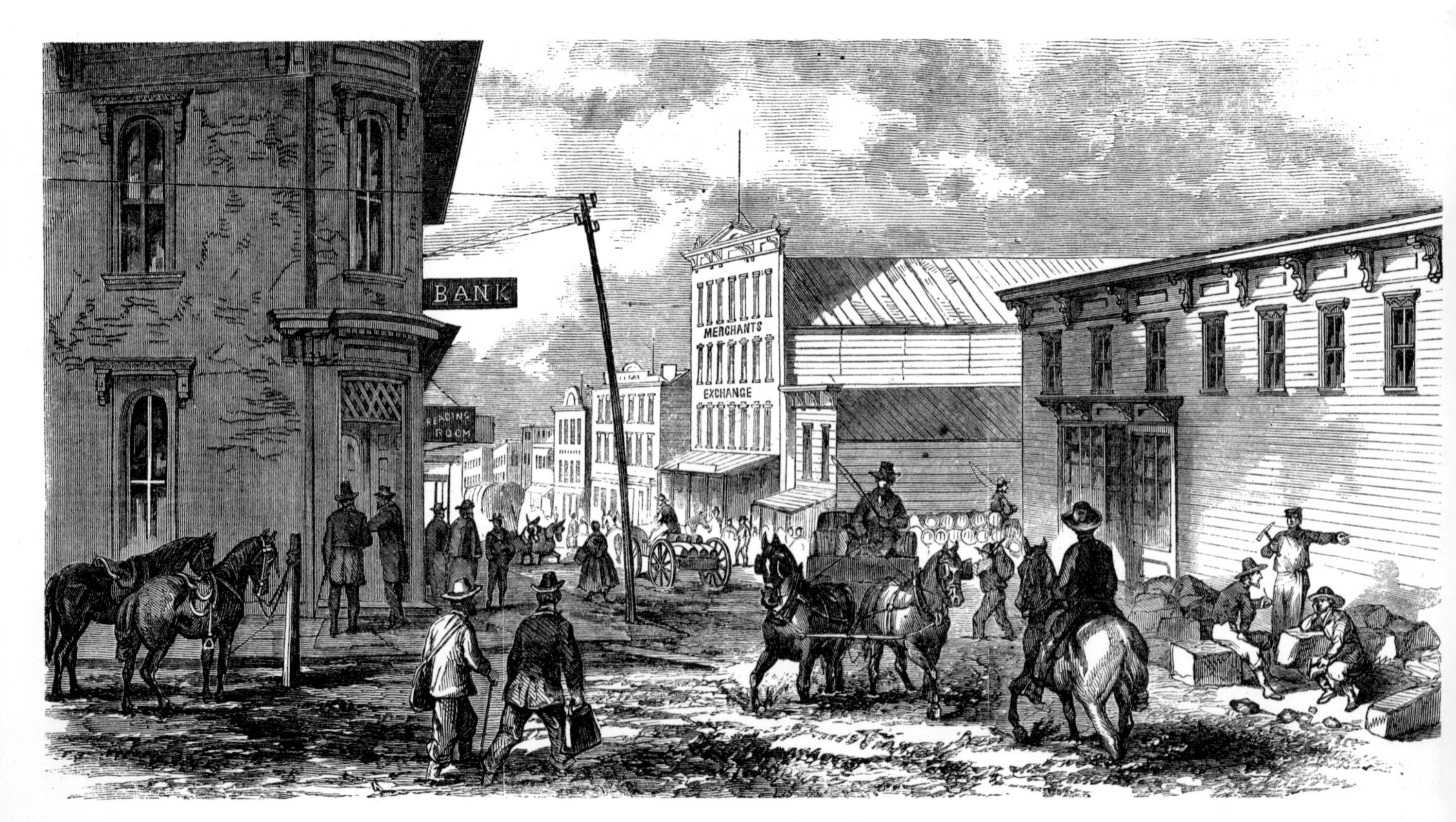

Hauling Oil at Titusville before 1866

First Tank Cars—Densmore Type

EMPIRE LINE.

The Empire Transportation Company offers to the business community a

RELIABLE FAST FREIGHT LINE

BETWEEN THE

East and the West!

AND THE

Great Oil Regions of Pennsylvania,

VIA THE

Philadelphia & Erie Railroad

AND ITS CONNECTIONS.

It owns and controls all the cars of its Line, which is managed by men of long experience in the business, and no effort will be spared on their part to render satisfaction to its patrons.

Parties ordering goods from the East will please direct shippers to mark packages EMPIRE LINE, and refer them to the Agents of the Company for shipping directions, etc., etc.

GEORGE W. RISTINE, General Freight Agent, Philadelphia, Pa.

GEORGE M. BALL, Eastern Superintendent, Philadelphia, Pa.

GEORGE W. CROSS, Western Superintendent, Cleveland, Ohio. 6jy 1y

Empire Transportation Company Advertisement, September 3, 1877.

Iron Boiler Tank Car, 1880

An improvement in the facilities for transporting oil was the appearance of the Empire Transportation Company in the oil region in the spring of 1865. It was organized by the Pennsylvania Railroad for the purpose of securing a larger share of the oil traffic originating along Oil Creek for the Philadelphia & Erie Railroad, which it then operated under a lease. About ten railroads connected with the Philadelphia & Erie to form direct routes to the east and west. A shipper using this route, however, had to make arrangements with each road. To overcome this inconvenience, the Empire Line acted as an intermediary between shippers and the various lines connecting with the P & E. It furnished its own cars and offered cheap, fast, and reliable service to the seaboard.

Samuel Van Syckel

From the moment that Van Syckel got the idea of transporting oil from Pithole to Miller Farm until he completed the project, he was the subject of ridicule. People would sarcastically inquire, "Do you intend to put a girdle around the world?" or "Can you make oil run uphill?" While his pipe line was being laid, it was maliciously cut in several places by teamsters who saw a menace to their monopoly and occupation. Along its entire length Van Syckel had to station guards to prevent interference with the pipe line.

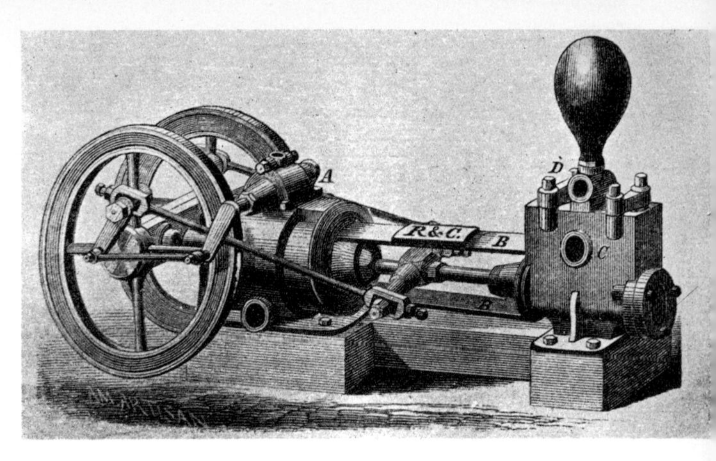

Reed and Cogswell Steam Pump. Three of these pumps were used by Van Syckel to force oil through his pipe line.

Pipe Line Dump—Pithole. None of the first pipe lines at Pithole connected directly to the tanks at the wells. Each line had dump tanks to which oil was hauled from the wells and dumped, then pumped away.

Miller Farm, Oil Creek, 1865-1868

South of Titusville on the Oil Creek Railroad, Miller Farm was the terminal for Van Syckel's pipe line from Pithole.

Henry Harley

In the spring of 1866 Henry Harley, a commission dealer
 oil, built two pipe lines from Benninghoff Run to Shaffer
rm a distance of two miles. Each pipe had a daily capacity
 1,500 to 2,000 barrels. Immediately, over 400 teamsters
ere thrown out of work and, as a result, Harley had to hire
atchmen to patrol the route and guard his storage tanks.

W. H. Abbott

Not long after the completion of Van Syckel's pipe line, he lost control of it to the First National Bank of Titusville. The line was sold to W. H. Abbott and Henry Harley.

Abbott & Harley Pipe Line Terminal, Shaffer Farm

ALLEGHENY TRANSPORTATION COMPANY OFFICE, MILLER FARM. In 1867 Abbott & Harley acquired control of the pipe line running from the Noble well to Shaffer farm and merged it with the Pithole and Benninghoff Run lines under the name of the Allegheny Transportation Company, the first great pipe-line company, with Harley as President. This company became closely associated with the Erie Railroad in its oil business.

PENNSYLVANIA TRANSPORTATION COMPANY OFFICE, MILLER FARM. Organized in 1871 under the leadership of Henry Harley, the Pennsylvania Transportation Company with a capital of $1,700,000 brought about a combination of the Oil Creek Railroad pipe-line interests and those of the Allegheny Transportation Company. With headquarters at Mille Farm, it controlled nearly 500 miles of pipe line tc Tidioute, Irvineton, Oil City, Shamburg, Pleasantville anc Titusville.

The Depot at Titusville

HE REED WELL, SMITH FARM, 1863. Although there had een several attempts to increase the production of oil wells y using torpedoes, William Reed claimed to be the inventor the torpedo. During the summer and fall of 1863, Reed ade some torpedoes using gunpowder and tried three of them on his well. These experiments were considered failures and Reed apparently abandoned the idea.

Later, in 1867, Reed applied for a patent but the Patent Office had already recognized the claim of Colonel E. A. L. Roberts.

Colonel E. A. L. Roberts

Endowed with an inventive ability, Colonel E. A. L. Roberts, while associated with his brother, W. B. Roberts, in the dental business in New York, made many improvements in dental science and dental operations. On the outbreak of the Civil War, he actively engaged in raising regiments for the Union and forwarding them to the fighting front. In 1862 Roberts was appointed Lieutenant Colonel of the 28th New Jersey Volunteers. In December of that year he conceived the idea of opening up the veins and crevices in oil-bearing rock by exploding an elongated shell or torpedo therein. Failing health compelled him to resign his command and return to his home in New York in 1864. Late in that year Roberts made six torpedoes and an application for a patent was filed. After some delay he started for Titusville to test them. Roberts' efforts to introduce his torpedo met with all kinds of rebuffs; oilmen were afraid that torpedoes would spoil their wells, but Roberts was persistent. In the latter part of January 1865, he obtained permission to explode his first torpedo in the Ladies well near Titusville, which greatly increased its production, created an immense interest in the torpedo, and demonstrated the success of his invention. From that day on, there was an enormous demand for the Roberts torpedo. Hundreds and thousands were used, and they added millions of dollars to the wealth of the oilmen. Four other individuals claimed patents for the same invention, so it was not until November 20, 1866 that a patent was issued to Roberts as the inventor of the torpedo.

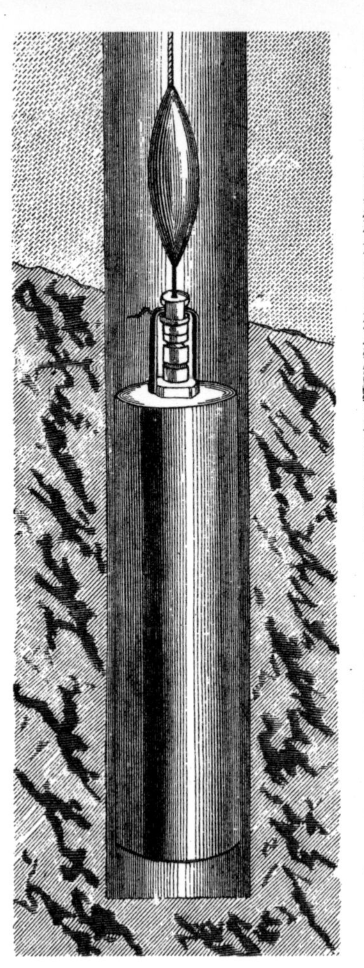

The Roberts Torpedo

This diagram from a contemporary history shows Roberts torpedo filled with gunpowder at the bottom of oil well. The one to the right shows the torpedo after iron weight had been dropped to explode the charge. Th torpedo, on being exploded, drove out the paraffin or oth coagulated matter and, at the same time, opened fissures in the solid rock. Nitroglycerine was later substituted f gunpowder.

The First Roberts Torpedo Factory was located on Hammon Run about half a mile south of Titusville.

W. B. Roberts, brother of E. A. L., furnished most of the noney used in developing and testing the torpedo and eceived a half interest in the enterprise. In the spring of 865, he organized the Roberts Petroleum Torpedo Company; the next year he was elected Secretary; and in 1867 e became President.

Oil producers, chafing because they considered the price f the torpedoes exorbitant, subscribed thousands of ollars to fight the monopoly. An army of "moonlighters"—en who put torpedoes in wells at night to avoid paying oberts his fee—came into existence. Roberts' agents combed e oil fields, gathered evidence, and hundreds of lawsuits ere started. The "torpedo war" became general and uncompromising. Finally, in June, 1880, the United States upreme Court upheld the Roberts patent and the producers ere forced to surrender.

W. B. Roberts

New York Feb 13th 1865

We the undersigned persons agree to purchase of E. A. L. Roberts his invention or improvements in enlarging the producing capacity of oil wells by the use of a Torpedo exploded in the bottom or at any other point of an oil well to open the veins by cracking the rock or blowing out the Paraffine or other substance that may clog or obstruct the oil. The sum to be paid by the said subscribers is to be $21,000 (twenty one thousand) divided into shares of $1000 each. The same to be put into a Stock Company of $200000 more or less at any time the majority of the subscribers may deem proper

Names of Subscribers			Amt
E. A. L. Roberts	superintendent	Ten Shares	$10,000
W. B. Roberts	treasurer	Two & one half	2,500
E. Titus Jr		Two & one half	2,500
E. Titus	President	Two	2000

Agreement to Organize the Roberts Torpedo Company

ROBERTS TORPEDO.

As litigation is much less than formerly, and hoping that it will entirely cease,

We Propose to Give the Producers the Benefit of the Following Schedule:

System of Orders for Roberts Torpedo

An order for $300 for $225............25 per cent. off
An order for $600 for $450............30 per cent. off.
An order for $1 200 for $800........23⅓ per cent. off.
An order for $2,400 for $1,500......37½ per cent. off.
An order for $5,000 for $2,750.......45 per cent. off.
An order for $10,000 for $5,000......50 per cent. off.

☞ Orders can be made up by two or more persons. All application for orders must be accompanied by the cash, and addressed to E. A. L. ROBERTS, Titusville, Pa., to receive prompt attention. No orders will be honored except signed by

E. A. L. Roberts, Titusville, Pa.

Roberts Torpedo Advertisement, September 3, 1877

Delivering a Shot to a Well

Pouring the Shot

Adam Cupler, Jr. & Company's wagon

Torpedoing oil wells was a hazardous business. Many shooters were killed either in hauling or handling the shot. Adam Cupler, Jr., of Titusville, one of the best known well shooters in Pennsylvania, started out early Monday morning, October 12, 1903, to deliver a shot to the Amos Klinger farm at Fagundas. While driving through East Titusville about 6:30 something set off the nitroglycerine. There was a terrific explosion; the wagon was blown to pieces; Cupler was horribly mangled and killed; but strange to say, his two horses escaped alive.

A Torpedoed Well

CHAPTER VII

BETWEEN OIL CREEK AND THE ALLEGHENY RIVE

AFTER the decline of Pithole, the tide of development moved in 1867 to the Tidioute and Triumph Hill region. For years prior to the drilling of Drake's well, there had been a small village at Tidioute, whose inhabitants engaged in the lumbering business. Among the most prosperous and prominent lumbermen were the Grandins—Samuel and his sons, John and E. B. When the news of Drake's well reached the village of Tidioute, John saddled his horse, galloped away, purchased the Campbell farm, below the village, and started drilling the first well in Warren County. While wells were drilled all around Tidioute, Triumph Hill, a short distance away, became the greatest center of excitement.

Late in 1867 there was a grand rush to upper Cherry Run. Two miles east of Miller Farm and four northwest of Pithole on the Stowell farm, the Pittsburgh & Cherry Run Petroleum Company under the able direction of Dr. G. S. Shamburg finished a well in February 1866. It was 800 feet deep and pumped 100 barrels a day. The depression prevented any extensive operations until December 1867, when the Fee and the Jack Brown wells on the Atkinson farm began flowing 400 barrels apiece. A lively town soon developed and wa called Shamburg as a compliment to the genial doctor. Th Tallman, Goss, Atkinson, Stowell and other farms increase the production to 3,000 barrels. The producing belt, how ever, was relatively small and the Shamburg wells soo declined. More important as a factor in the decline of Sham burg was the excitement which had developed at Pleasan ville, for it was the usual thing to follow the crowd from on big strike to another.

Though a small well, the Harmonial, drilled in the fa of 1867, opened up the Pleasantville field and started wh promised to be a second Pithole. Shortly, at least fifty wel were being drilled and by July 1868, the production wa more than 600 barrels. For the next month or six week hundreds of men flocked to Pleasantville. From dayligh until dark the streets of the borough were thronged. By th end of August the average daily production had increase to over 2,000 barrels. The wells soon became exhausted, th crowd began deserting Pleasantville, and by the end of 18 the boom had collapsed.

THE ECONOMITE WELLS, TIDIOUTE, 1864

A few years prior to the drilli of the Drake well, the Econom Society, a group of German Pietis came into possession of some six th sand acres of land opposite Tidiou During the summer of 1860 oil w found on the land, which created gr excitement. Drilling began imme ately and the property was rapidly c veloped. A large boarding house w erected for the workmen, wh speech and manners were regulated printed rules. In March 1862, t society had four good producing we During 1868 the society sold ab 100,000 barrels of oil. Seventy-six we were drilled between 1860 and 18 the largest one produced 250 barr a day. The Economites never pump their wells on Sunday.

Tidioute, 1864

Train and Muddy Road, Tidioute

Empire Line Office, Tidioute

Oil Barges, Tidioute

Barrel Yard and Exchange Hotel, Tidioute

Wells on Dingley Run near Tidioute

The Triumph Oil Company of Pittsburgh purchased th W. W. Wallace farm on Triumph Hill in 1864. During th summer of 1867 its daily production was 2,000 barrels.

THE KING BEE WELL, TRIUMPH HILL, was owned by S. Hughes & Davis.

TRIUMPH HILL

For five years, Triumph Hill, the highest elevation in the vicinity of Tidioute, was busy and prosperous yielding hundreds of thousands of barrels of oil and advancing Tidioute to a town with 5,000 population. The east side of the hill was a forest of derricks, crowded like trees in a grove. Over the summit and down the west side, one observed the same development. Wells nine hundred feet deep pierced sixty feet of oil-bearing sand and produced the purest oil in the region. No other field had so many derricks upon the same area. Amid drilling wells, pumping wells, oil tanks, and engine houses on the summit, the town of Triumph came into existence. Drilling wells too close together, flooding, and the opening of a new field at Pleasantville were responsible for scores of wells being abandoned at Triumph Hill.

Dr. G. S. Shamburg's Home, 1867

Developments in 1866 and 1867 resulted in the striking of a number of wells on the lands of the Pittsburgh & Cherry Run Petroleum Company, the Shamburg Petroleum Company, and other companies at the headwaters of upper Cherry Run. Operators in this field used the best type of machinery, and it was radically different from that previously used. Instead of derricks 20 to 30 feet high, they built them 56 feet high. In earlier years the weight of drilling tools was about 600 pounds; the Shamburg operators used tools weighing 1,600 to 1,800 pounds. Large casing was also used. Operations in the Shamburg pool were almost invariably profitable, and handsome fortunes were realized.

As usual in the oil region, a town developed around the first wells drilled by Dr. Shamburg; it was commonly known as Shamburg. At its height Shamburg had a population of about 4,000. Because of the Fee well and the

General View of Wells, Shamburg

Jack Brown well, another village, Atkinson, developed in the district. Between Shamburg and Atkinson, Backus City sprang into existence.

H. Spears Well, Shamburg Petroleum Company, started flowing in July 1868 at 450 barrels a day. All of the wells drilled on this tract produced oil. The farm originally belonged to Marshall Goss.

The Atkinson farm adjoined the Goss farm on the east. The striking of the Jack Brown well in December 1867, and the Fee well in January 1868, which flowed about 400 barrels apiece, precipitated the rush to Shamburg. Fee well No. 2, drilled in April 1868, yielded about 150 barrels a day. Fee well No. 3, drilled in the following June, flowed 1,500 barrels, then fell off to 500 and less. These four wells were the most productive wells on the farm.

The Fee Well, Atkinson Farm, Shamburg

Lady Stewart Well, Shamburg, reported to be the largest producing well in the oil region in 1869, yielding 180 barrels a day.

The Tallman farm was owned by F. W. Andrews, F. L. Backus, Lyman Stewart, Milton Stewart, C. H. and W. C. Andrews. It lay north of the Atkinson farm. Commencing in September 1867, the farm had produced by January 1, 1869, over 200,000 barrels of oil, which sold for nearly $800,000.

Beardsley House, Shamburg

Lyman Stewart's Home, Shamburg

Like many another Pennsylvania oilman, Lyman Stewart moved to California in the early eighties. He and W. L. Hardison formed a partnership which became one of the leading oil firms on the Pacific Coast. Later, Stewart served for many years as President of the Union Oil Company.

SHAMBURG, 1868

Post Office, Shamburg

Hardware Store, Shamburg

Red Hot near Shamburg

The Windsor Brothers of Oil City drilled a well a short distance from Shamburg in 1869, which flowed 350 barrels a day. Other operators rushed in to drill and the town of Red Hot came into existence. The territory lacked staying qualities and by 1871 the town had practically disappeared.

Stores at Red Hot

Travelers Rest, Red Hot

HARMONIAL WELL, PLEASANTVILLE

One day, late in the fall of 1867, Abram James, an ardent spiritualist, was driving from Pithole to Titusville with three friends. A mile south of Pleasantville the "spirit guide" caused him to jump out of the conveyance and leap over the fence into a field on the William Porter farm. Hurrying to the north end of the field, James fell to the ground, marked the spot with his finger, thrust a penny into the dirt and fell back pale and rigid. Restored to consciousness, James told his friends of a revelation that streams of oil lay beneath the soil. He leased the property, borrowed money, raised a derrick over the spot where the penny lay, and commenced drilling amid the scoffs of unbelievers. When down 700 feet and past the third sand rock, he became the laughing stock of the region; but "Crazy James" kept on drilling. When he proceeded to build tanks to receive the expected oil, people laughed louder. Early in February 1868, James struck oil and his well, called the Harmonial in honor of the spiritual philosophy, pumped 130 barrels a day. The usual hurly-burly followed. Operators at once rushed in to secure leases on adjoining farms, and the fact that experienced oilmen were willing to pay $500 to $1,250 an acre created confidence in the territory. New strikes increased the excitement.

GENERAL VIEW, WILLIAM PORTER FARM, PLEASANTVILLE

The first producing farm in the Pleasantville field was the one on which the Harmonial was drilled. Though no one put any faith in the highly supernatural narrative of James, every operator located his well as near to the Harmonial as he could get. In a short time scores of wells were being drilled, and they turned out to be good wells. It came to be an established fact that every well drilled was likely to prove a paying one. By the spring of 1868 the oil scenes of Pithole were being reenacted at Pleasantville.

Wells on the Armstrong Farm, Pleasantville, located about three-fourths of a mile southwest of Pleasantville. Included amon
the men in this picture is a group of Englishmen who were visiting the oil region.

In two months' time during th
summer of 1868, the population o
Pleasantville jumped from 1,000 t
3,000 and what had been a quiet an
peaceful little borough was trans
formed into a bustling oil town. Grea
rows of shanties and large fram
buildings began to appear among th
neat white houses, making a curiou
mixture of the old and the new. Ever
road leading to the town was crowde
with travelers, while the few hotel
afforded scarcely standing room fo
the crowd.

Oilmen and Drilling Tools, Pleasantville

"The Corners," Pleasantville

During the memorable summer of 1868 the number of isitors who traveled back and forth to Pleasantville from itusville and the surrounding towns averaged 1,000 persons day. Pleasantville was red hot! It possessed many characteristics of pioneer oil towns but none of the repulsive features. An established civilization preceded the oil rush and the high moral standards of the residents kept the community relatively free from vice of all sorts.

Chase House, Pleasantville
Instead of erecting new buildings, people moved many of those at Pithole, like the Chase House, to Pleasantville.

Brown Brothers Store, Pleasantville. John Brown cam[e] from New York City to Pleasantville in 1833 and opened [a] general store. He died in 1861, leaving a very prosperou[s] business to his three sons—John F., Samuel Q., and Alex. W[.] Through their business sagacity, they amassed a fortun[e] from the merchandising business; they carried a $100,00[0] stock in the store and their annual trade amounted to abou[t] $200,000. Their fortunes were greatly increased through o[il] investments. Samuel Q. Brown later became President o[f] the Tidewater Pipe Company.

Ned Pitcher Well, Pleasantville

First Large Pumping Station near Pleasantville. Located about two and a half miles from Pleasantville at the Nationa[l] Wells, this pumping station belonged to the Titusville and Tidioute Pipe Line Company.

CHAPTER VIII

TITUSVILLE THE QUEEN CITY

Titusville possessed numerous advantages over all the other towns in the oil region. Located in a broad valley with room for expansion, it was already a village of several hundred people with certain social institutions when Drake struck oil. Being the nearest town to the site of the Drake well, Titusville received a tremendous impetus from the rush to Oil Creek and, as new fields were opened, it benefited immeasurably. The completion of the Oil Creek Railroad to Titusville in the fall of 1862 increased its commercial importance; it became not only the leading shipping center but the most convenient approach to the oil field for thousands of people. From the beginning of the oil boom, Titusville's leading citizens struggled to keep the village a peaceful and law-abiding community, a decent place in which to reside—and they did. This induced many an oilman to bring his family and make Titusville his home.

As a result of these influences, Titusville emerged in 1870 as the commercial, financial, and cultural capital of the oil region with a population of ten to twelve thousand. It was far ahead of other towns in factories and industries connected with oil production. There were four dry-goods stores whose individual business amounted to $100,000 to $125,000 a year. Deposits in two banks exceeded $1,000,000, while the daily transactions at these banks averaged $75,000. Numerous oil companies made Titusville their headquarters. The Titusville Oil Exchange, organized in 1871, flourished and exerted a profound influence upon the marketing of oil.

The absorption in material things and the eagerness to acquire wealth did not cause the citizens to forget the cultural and religious side of life. Fine churches were built, a private academy was established, and a men's literary society met regularly. Many of the leading lecturers of the day spoke before capacity audiences. Fully as brilliant as the lecture talent were the musicians who gave concerts.

In view of these distinguishing characteristics, it isn't strange that Titusville came to be widely known as the "Queen City."

Titusville from South Hill, 1873

The Mansion House about 1874

West Spring Street and the Parshall Hotel about 1874

By 1870 most of the streets had been graded and drained. New plank walks extended in every direction and street lamps lighted by gas had been located all over town. Brick buildings were beginning to replace the hastily-built wooden structures of the early day. The Parshall Hotel, built by the wealthy James Parshall, opened in December 1870, overshadowed all other buildings in Titusville. It was a magnificent structure furnished in rich and elegant style. An opera house, with a seating capacity of 1,500, occupied the second floor. Musicians, like Clara Louise Kellogg, Christine Nilsson, Ole Bull, and Theodore Thomas and his orchestra, performed here. Bishop Simpson, John B. Gough, and other distinguished lecturers spoke in this opera house. President Grant addressed briefly an assembled throng from one of the balconies of this hotel on his visit to the oil region in September 1871.

Granger & Company Store, 1868

Though occupying a modest frame building at Franklin Street and Central Avenue, Granger & Company did an enormous amount of grocery business. Its stock seldom fell below $100,000. Selling flour was a specialty and the firm sold on an average over one hundred barrels a day.

E. K. Thompson's Drug & Chemical Depot, 1867

Henry Harley invited the members of the Titusville ommercial Club to visit New York for a week as his guests. n January 17, 1871 a special car on the Oil Creek Railad carried the twenty-one members to the city. They were reeted by Jim Fisk, attended the opera, had a magnificent anquet at the Hoffman House, went to the horse races, nd in general had a glorious time. In order to express their ppreciation for what Harley had done, the club decided to nder him a banquet at the Parshall Hotel on May 24 that ould include the elite of the oil region. It was the most shionable gathering ever held in Titusville up to that time. he banquet room of the Parshall Hotel was a scene of egance. Everything that wealth could do was done to ake the occasion a memorable one. As a token of their steem of "Pipe Line Harley," the club presented him with a erling silver cup costing $3,700. On the exterior were enraved scenes from the oil region.

John J. Carter

John J. Carter located in Titusville in the summer of 1865—immediately after being mustered out of the Union Army—and went into the men's clothing business, which he maintained until 1877. At that time, Carter sold out and began a remarkable and unique career in the oil business.

Stettheimer's Clothing Store

Carter's Clothing Store about 1869

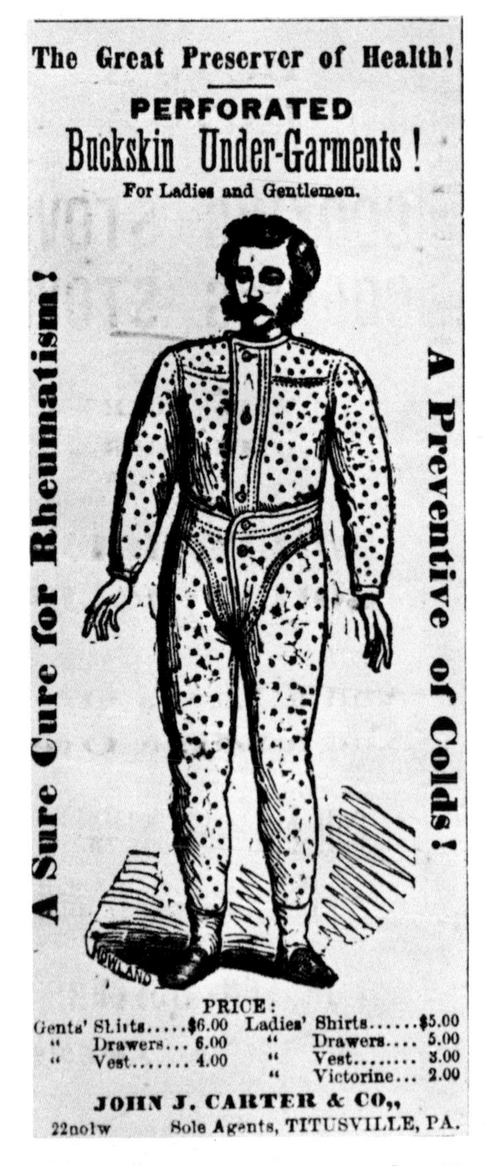

The Great Preserver of Health!

PERFORATED

Buckskin Under-Garments!

For Ladies and Gentlemen.

A Sure Cure for Rheumatism!

A Preventive of Colds!

PRICE:

Gents' Shirts	$6.00	Ladies' Shirts	$5.00
" Drawers	6.00	" Drawers	5.00
" Vest	4.00	" Vest	3.00
		" Victorine	2.00

JOHN J. CARTER & CO.,

22no1w Sole Agents, TITUSVILLE, PA.

Carter's Advertisement, November 23, 1870

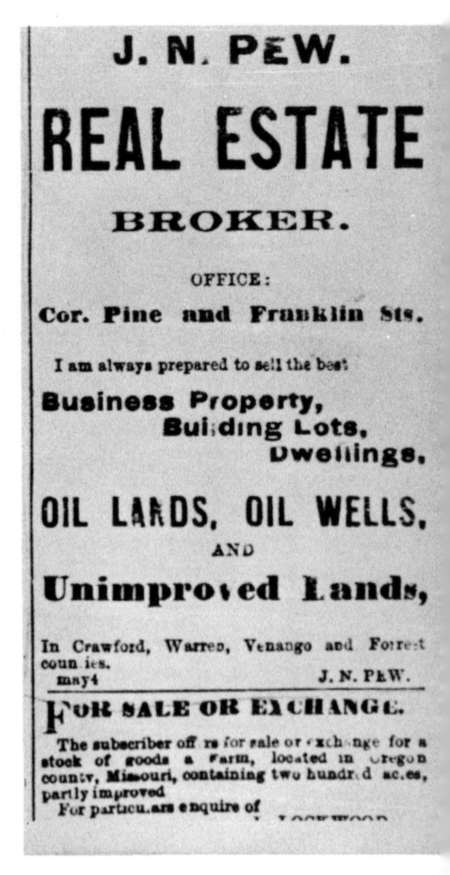

J. N. PEW.

REAL ESTATE

BROKER.

OFFICE:

Cor. Pine and Franklin Sts.

I am always prepared to sell the best

Business Property,
Building Lots,
Dwellings,

OIL LANDS, OIL WELLS,

AND

Unimproved Lands,

In Crawford, Warren, Venango and Forrest counties.

may4 J. N. PEW.

FOR SALE OR EXCHANGE.

The subscriber offers for sale or exchange for a stock of goods a farm, located in Oregon county, Missouri, containing two hundred acres, partly improved

For particulars enquire of

J. N. Pew's Advertisement, June 3, 1872

J. N. Pew, the father of J. N. Pew, J and J. Howard Pew of the Sun C Company, lived in Titusville for number of years.

Although petroleum was used exclusively in the early years as an illuminant, various efforts were made to use it for other purposes. About 1877 Professor Charles J. Eames discovered a process by which iron could be made through the use of oil as fuel. Consequently, a building was erected in Titusville in 1879 for making iron by this process. By September the plant had been completed and the manufacture of iron started. The novel spectacle of petroleum heating a puddling furnace and providing heat for all other purposes attracted great crowds to the plant each day.

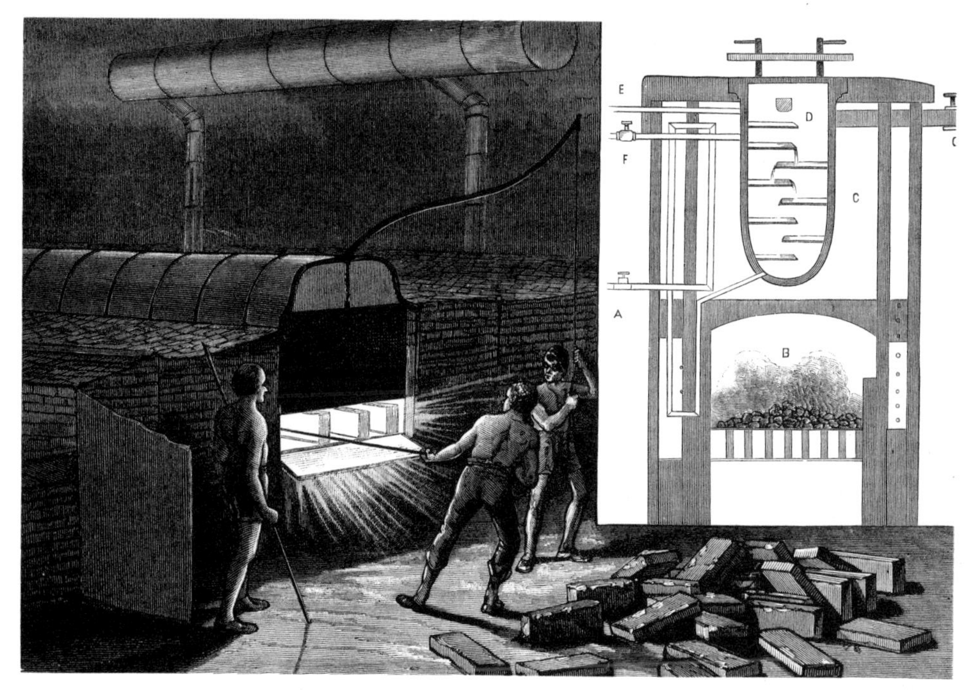

Eames Petroleum Iron Works, Titusville, 1879

Excursion Train to Oil Creek Lake

Special trains often ran during the summer from Titusville to Oil Creek Lake, now Canadohta Lake, where oilmen and their families relaxed and enjoyed a day's outing.

In 1872 eleven refineries were located in Titusville. The Acme Oil Company, affiliated with Standard Oil, commenced business in Titusville in 1875 by the purchase of the Porter, Moreland & Company refinery, which became Acme No. 1 and the purchase of the Bennett, Warner & Company, re-named Acme No. 2. In 1876 the Octave Oil Company's refinery and the John Jackson refinery were purchased, consolidated, and named No. 3.

Acme Refinery No. 3, and Workmen

TITUSVILLE
1884

Heisman's Cooper Shop

Barnsdall Oil Company Office. Many of the oil compa-nies, like the Barnsdall, had offices in Titusville.

Early dealers in petroleum lacked correct informatio regarding the amount of oil produced, a necessary facto in determining the price. Moreover, buyers seldom quote the same price to producers at widely separated points. N one seemed to know or care about actual market conditions each selling or buying to suit his own particular fancy. A third handicap in ascertaining the real market value of o was the fact that most transactions in the oil region wer unknown; as they were not a matter of record no correc quotation could be given.

J. M. Henderson Oil Brokerage Office and Oilmen

Titusville Oil Exchange

Under the circumstances there was a definite need for an rganization where buyers and sellers could meet, ascertain vith some accuracy the amount of oil actually being pro-luced and sold and transact their business in accordance vith certain rules and regulations. In January 1871, one ıundred and sixteen dealers, producers, refiners, and brokers .long Oil Creek organized the first oil exchange, the Titus-ille Oil Exchange. L. H. Smith was elected President and ohn D. Archbold, Secretary.

As the Titusville Oil Exchange increased in membership and business, there was considerable agitation for erecting a building—one that would do honor to Titusville and compare favorably with those in Oil City and Bradford. In 1880 preliminary arrangements were made, stock subscribed, a site selected, and construction started. On January 27, 1881, with hundreds of oilmen present, the new exchange building was dedicated and opened for business.

The New Oil Exchange, West Spring Street

Oilmen in Front of the Oil Exchange Building

St. James Memorial Church and Appleton's Collegiate Institute

Prior to 1859 there were three religious denominations in Titusville—the Presbyterians, Methodists, and Universalists. During the period of Titusville's rapid growth, from 1860 to 1865, five more congregations were formed. St. James Memorial Church was organized as a Protestant Episcopa mission in June 1863. Through the efforts of Colonel Drake W. H. Abbott, George M. Mowbray, and other early friend of the church, a handsome house of worship was built.

Believing that Titusville should have something beyonc the grades of the common school, a group of leading citizen founded Appleton's Collegiate Institute in 1865. It openec under the supervision of a committee of the vestry of th Episcopal Church, but it was non-denominational. Th building nearest the reader is where the school opened The population of Titusville grew so rapidly that the de mand for school facilities could scarcely be met.

F. S. Tarbell's Residence, East Main Street

In 1870 F. S. Tarbell moved his family from Rousevill to Titusville. The new house into which they moved ha previously been the Bonta House at Pithole. Built at a original cost of $80,000, Tarbell paid $600 for it, tore dow the building, and used the material to build this home i Titusville. Ida M. Tarbell, his daughter, thirteen years ol at this time, graduated from the Titusville High School i 1876.

Drake Street School

In 1871 the school directors erected ıis large brick schoolhouse at the ortheast corner of Walnut and Drake treets.

Eighth Grade Class, Drake Street School, June 1881

Celebrating the Centennial Exposition, E. O. Emerson's Residence

Mrs. George R. Harley and Mrs. J. H. Cogswell

George K. Anderson, whose income was estimated a
$5,000 a day for two years, built this stately mansion. H
owned thousands of acres of land and, at one time, he i
said to have carried $315,000 in life insurance. Later, th
residence was purchased by E. O. Emerson, another oi
prince. Emerson's house and grounds, with a grand foun
tain, velvet lawns, smooth walks, tropical plants, and al
kinds of flowers, were admired by all. It was a most popula
place for hundreds of citizens on summer evenings. Titus
villians celebrated the Centennial Exposition during th
summer of 1876 on the decorated lawn of Emerson's home

Mrs. John J. Carter

With the firm establishment of 'itusville as the metropolis of the oil :gion, oilmen and others decided that might be possible to go farther and o worse than make their homes in itusville. The result was that Titus-ille became a most attractive resi-ential center. In addition to the naller dwellings, many large and ately houses costing thousands of ollars were erected.

The Cady house cost $35,000 to uild. D. H. Cady who made a ortune at Pioneer and Shamburg, urchased the house and furnished it xtravagantly. It was purchased by ohn D. Archbold in 1877; later it erved as the home of John J. Carter.

Charles Gibbs Carter in 1884

Many of the young men in the)'s, like Colonel Carter's son, received more formal education than their athers, going to schools like Andover nd Yale. Later, most of them became rominent figures in the business and rofessional worlds.

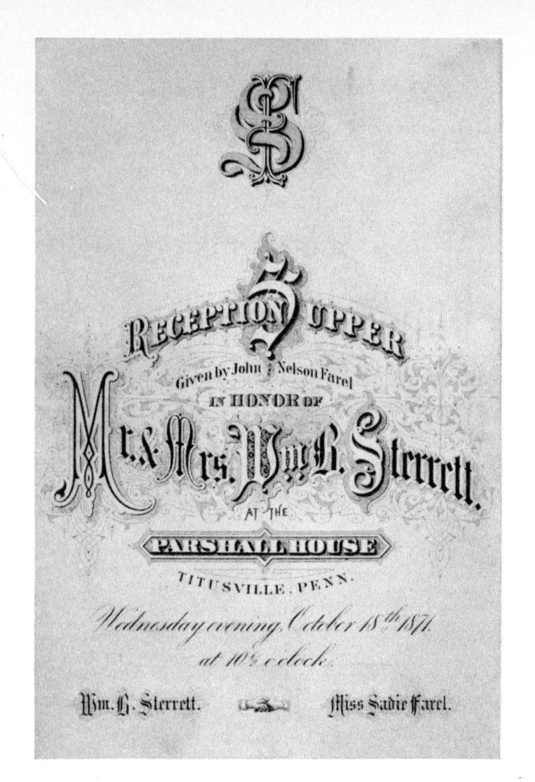

Reception Supper

Given by John & Nelson Farel

in honor of

Mr. & Mrs. Wm. B. Sterrett.

at the

Parshall House

Titusville, Penn.

Wednesday evening, October 18th 1871.

at 10½ o'clock.

Wm. B. Sterrett. Miss Sadie Farel.

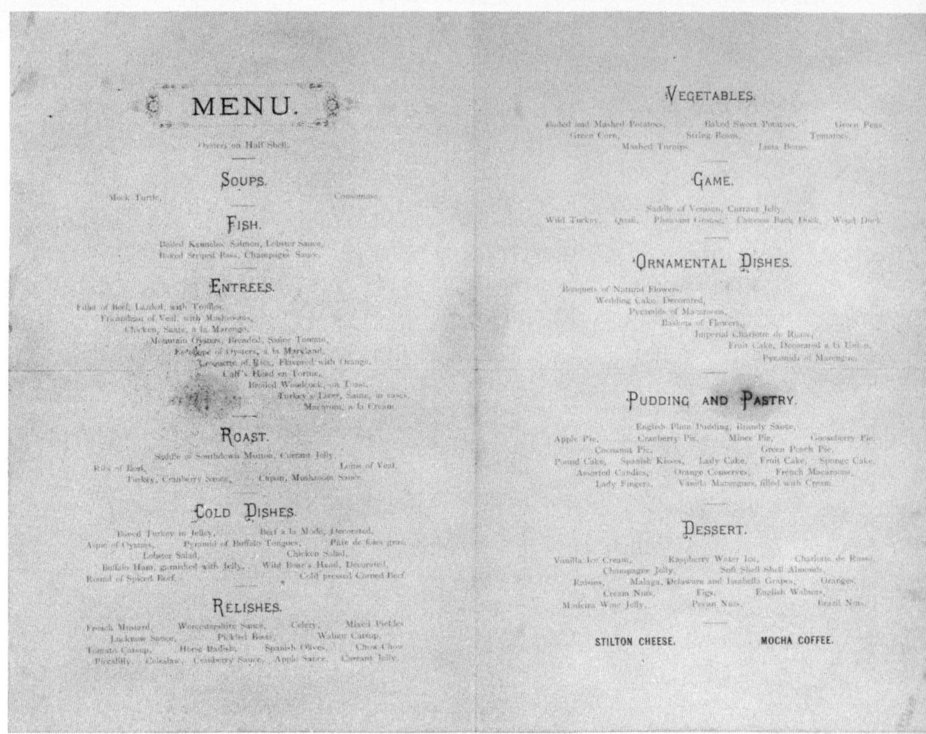

MENU.

Oysters on Half Shell.

SOUPS.

Mock Turtle, Consomme.

FISH.

Boiled Kennebec Salmon, Lobster Sauce,
Baked Striped Bass, Champagne Sauce.

ENTREES.

Fillet of Beef, Larded, with Truffles,
Fricandeau of Veal, with Mushrooms,
Chicken, Saute, a la Marengo,
Mountain Oysters, Breaded, Sauce Tomato,
Escallops of Oysters, a la Maryland,
Croquette of Rice, Flavored with Orange,
Calf's Head en Tortue,
Broiled Woodcock, on Toast,
Turkey's Liver, Saute, in cases,
Macaroni, a la Cream.

ROAST.

Saddle of Southdown Mutton, Currant Jelly,
Ribs of Beef, Loins of Veal,
Turkey, Cranberry Sauce, Capon, Mushroom Sauce.

COLD DISHES.

Boned Turkey in Jelley, Beef a la Mode, Decorated,
Aspic of Oysters, Pyramid of Buffalo Tongues, Pâte de foies gras,
Lobster Salad, Chicken Salad,
Buffalo Ham, garnished with Jelly, Wild Boar's Head, Decorated,
Round of Spiced Beef, Cold pressed Corned Beef.

RELISHES.

French Mustard, Worcestershire Sauce, Celery, Mixed Pickles,
Lucknow Sauce, Pickled Beets, Walnut Catsup,
Tomato Catsup, Horse Radish, Spanish Olives, Chow Chow,
Piccalilly, Coleslaw, Cranberry Sauce, Apple Sauce, Currant Jelly.

VEGETABLES.

Boiled and Mashed Potatoes, Baked Sweet Potatoes, Green Peas,
Green Corn, String Beans, Tomatoes,
Mashed Turnips, Lima Beans.

GAME.

Saddle of Venison, Currant Jelly,
Wild Turkey, Quail, Pheasant Grouse, Canvass Back Duck, Wood Duck.

ORNAMENTAL DISHES.

Bouquets of Natural Flowers,
Wedding Cake, Decorated,
Pyramids of Macaroons,
Baskets of Flowers,
Imperial Charlotte de Russe,
Fruit Cake, Decorated a la Union,
Pyramids of Marengue.

PUDDING AND PASTRY.

English Plum Pudding, Brandy Sauce,
Apple Pie, Cranberry Pie, Mince Pie, Gooseberry Pie,
Cocoanut Pie, Green Peach Pie,
Pound Cake, Spanish Kisses, Lady Cake, Fruit Cake, Sponge Cake,
Assorted Candies, Orange Conserves, French Macaroons,
Lady Fingers, Vanilla Marengues, filled with Cream.

DESSERT.

Vanilla Ice Cream, Raspberry Water Ice, Charlotte de Russe,
Champagne Jelly, Soft Shell Shell Almonds,
Raisins, Malaga, Delaware and Isabella Grapes, Oranges,
Cream Nuts, Figs, English Walnuts,
Madeira Wine Jelly, Pecan Nuts, Brazil Nuts.

STILTON CHEESE. MOCHA COFFEE.

Menu, Reception Supper, Sterrett-Farrel Wedding

The greatest social event in Titusville or in the entire oil region was the wedding of Miss Sadie Farrel and William B. Sterrett on October 18, 1871. It illustrates the brilliant character of the petroleum aristocracy which had developed in Titusville. The number of prominent guests, the elegance of the ladies' dresses, the numerous and costly presents, and the reception supper far exceeded anything ever seen in the oil region. At the conclusion of the ceremony in the First Presbyterian Church, a reception was held at the Parshall Hotel. To the irresistible strains of gay waltzes, galops, and polkas played by an orchestra from Buffalo, the guests danced until about 10:30, when 300 people sat down to a sumptuous wedding supper. Among the gifts to the bride were the following: from George K. Anderson a combination set of thirty solid silver spoons; from W. H. Abbott a combination berry set of fifteen pieces gold lined; from John and Nelson Farrel eighty pieces of solid silver; from R. H. Sterrett an elegant French clock and from Gibbs, Russell & Sterrett a solid silver tea set of six pieces.

J. C. McKinney's Residence

After operating lumber yards at Oil City and Franklin, J. C. McKinney began his career as an oil producer at Foster, below Franklin, in 1864. Forming a partnership with his older brother, John L., the men became leading operators in all the Pennsylvania fields. J. C. moved to Titusville in 1877, bought the Windsor mansion which he remodelled into one of the finest residences in the oil region.

CHAPTER IX

OIL DEVELOPMENTS 1870 TO 1885

)IL operators never ceased looking for new places to drill. ˄s the excitement along Oil Creek, at Pithole, Shamburg, 'leasantville, and Tidioute declined, the more daring pro-lucers groped their way down the Allegheny River below)il City and began wildcatting in Clarion, Armstrong, and utler counties. Rewarded with good producing wells, other ilmen followed. Foxburg, St. Petersburg, Karns City, 'etrolia, Millerstown, Parker's Landing and other places ecame new centers of excitement. For five years, from 870 to 1875, the Lower Region, as this area was called, ompletely overshadowed all other fields.

Concentration upon the production of oil was suddenly nterrupted early in 1872 by the news that the South Im-rovement Company, composed of the five leading refiners nd the railroads of the oil region, had been secretly rganized for the purpose of controlling oil refining and il transportation. Immediately all oilmen outside the ombine prepared to fight. They agreed not to sell any oil the ring. After a blockade lasting forty days, the gigantic nonopoly was defeated and destroyed.

The greatest event of the period was the opening of the mous field at Bradford. In the sixties Job Moses, living on a large tract near Limestone, New York, drilled some shallow wells and demonstrated the existence of oil in this area. Others in the vicinity drilled with indifferent success. During the summer of 1875 Frederick Crocker of Titusville drilled a well near the village of Bradford, eight miles south of Moses' wells. The well flowed over one hundred barrels a day and helped to precipitate a rush to Bradford. Energetic wildcatters pushed the drill in every direction. Drilling was easy and production increased as it never had before. In 1878 the daily production exceeded 6,000 barrels and in 1880, 50,000. Thousands of wells yielding twenty-five to two hundred barrels covered the hills and valleys around Bradford. By 1881 the Bradford field had been pretty thoroughly defined and production reached its peak—nearly 100,000 barrels a day! In other words, Bradford was producing nearly 83 per cent of all the crude in the United States and nearly 77 per cent of the world's oil supply.

The largest gushers of the time—the Armstrong, Phillips, and Christie wells—were drilled on Thorn Creek southwest of Butler in 1884. However, Thorn Creek, like Pithole, had a brilliant opening, a brief but dazzling career, and an inglorious end.

James B. Kerr Farm, Church Run, near Titusville, 1870.

hurch Run entered Titusville from the north and flowed rough the town. Wells were drilled up the run early in the xties but extensive operations were not undertaken until about 1870. The wells on Church Run were not large individual producers but they produced a great deal of oil; it was good quality oil and refiners sought it eagerly.

John D. Rockefeller in 1872

Rockefeller first entered the oil refining business in 1863 with the firm of Andrews, Clark & Company of Cleveland. By 1868 he was head of the largest refining business in the world—Rockefeller, Andrews & Flagler. The name of the firm was changed in January 1870, to the Standard Oil Company of Ohio.

An Oil Riot, 1872

Peter H. Watson

President of the South Improv- ment Company.

John D. Archbold in 1872

In 1872 John D. Archbold, one the youngest refiners in the oil regio lived in Titusville, and was one of th most active leaders in breaking up th South Improvement Company. Late he became President of the Standa Oil Company.

News of the formation of the Sou Improvement Company created a te rific storm in the oil region. In spite the excitement and tense feeling, the were only two or three instances mob violence. In April 1872, it w learned that a train of tank cars Rynd farm was loading with oil f Standard. A crowd gathered and pr vented the train from pulling out.

CYRUS D. ANGELL

In the early days of the petroleum industry there was no scientific method by which deposits of oil could be located. Theories, however, were not lacking. The most valuable of these was that held by C. D. Angell. After drilling several wells below Franklin and making some observations, Angell concluded that petroleum existed in "belts" or courses. He adopted the theory that two "belts" existed, one running from Petroleum Centre to Scrubgrass and the other from St. Petersburg to Butler County. While the mass of producers discounted the theory, they were nevertheless influenced by Angell's observations, which had much to do with the opening up of new territory in Clarion, Armstrong, and Butler counties.

The Mutual Pipe Line Company, with headquarters at Foxburg, furnished pipe-line service for the new developments in Clarion County.

When drillers moved down the Allegheny River, Clarion 'ounty was the first to attract attention and Foxburg became ıe jumping off place for St. Petersburg, Antwerp, Turkey ·ity, Dogtown, Triangle, Slambang, and Jefferson City. ·rokers, dealers, producers, and pipe-line men congregated ·ere. Dom Pedro, the Emperor of Brazil, President Ruther-ɔrd B. Hayes, and other distinguished persons visited Fox-·urg in its heyday.

FOXBURG IN 1872

Parker's Landing, 1874. Situated on the west bank of the Allegheny River in Armstrong County and below Foxburg, Parker's Landing became the capital of the oil trade for the lower oil field. Operators of every class and condition, speculators, gamblers, and adventurers thronged its streets. It had, according to reports, the vim and vigor of Oil City Rouseville, Petroleum Centre, and Pithole done up in single package. When incorporated in 1873, the population was about 4,000. With the rise of Bradford in the north Parker's sun began to set.

Brokers and Oilmen, Parker's Landing

LIZZIE TOPPLING
ench Kate's" Love-Rival, Whose Siren Wiles Lured Many
Iale Admirers Into the Land of Passionate Enchantment!

FRENCH KATE.

Ben Hogan's Floating Palace, Parker's Landing

When Parker was a red-hot oil town, Hogan opened a free-and-easy but a reform wave swept over the place and Hogan was forced to close. Not to be outdone, he bought a boat known as the "Floating Palace," put his gambling tables and girls aboard, and anchored it in the middle of the Allegheny River opposite Parker just outside the jurisdiction of the county authorities. Here Hogan continued his business with profit. Hogan was at Parker three years and made $210,000.

On Board Hogan's Floating Palace

PETROLIA, BUTLER COUNTY

Dimick, Nesbit & Company finished a wildcat well on April 17, 1872, on the line dividing the Jameson and Blaney farms about seven miles southwest of Parker. This was the noted Fanny Jane, which pumped 100 barrels a day an gave birth to the town of Petrolia. The Fanny Jane stirre the blood of the oil clan and operators began to arrive i May; by July, 2,000 people made their homes in Petrolia. charter was obtained and Dimick was chosen burgess at th first election in 1873. Fisher brothers paid $60,000 for th Blaney farm, and wells multiplied in all directions. Be Hogan opened his usual place of business. Petrolia pro pered for four years and then the boom collapsed.

Fairview Pipe Line Company Office, Petrolia. Owned by Vandergrift & Foreman of Oil City, the Fairview line was 125 miles long and connected Sheakley, Petrolia, Greece City, Millerstown, and Modoc.

Main Office, United Pipe Lines, Petrolia

KARNS CITY, BUTLER COUNTY

Success at Petrolia induced operators to move in a outherly direction. Stephen Duncan Karns leased an bandoned well a mile below Parker; he drilled through the and and the well produced twenty-five barrels a day. This ettled the question of oil being south of Parker. "Dunc" rilled on Bear Creek, secured the famous Stonehouse farm f 300 acres and in 1872 enjoyed an income of $5,000 a day. A mile south of Petrolia, on the McClymonds farm, Karns paid $8,000 for a well which the drillers were about to abandon; he resumed drilling and soon the well flowed about 100 barrels a day. The town of Karns City blossomed into a community of 2,500 people. For a year or two Karns was the largest producer in the oil region.

MILLERSTOWN, BUTLER COUNTY

Wildcatters moved on south from Karns City. Each suc-cessful well strengthened the "belt" theory. In April 1873, Shreve & Kingsley drilled on the Stewart farm, found a good sand, and the well flowed 140 barrels. The Shreve well started a stampede toward Millerstown, a quiet hamlet half a mile southeast of the well. "On to Millerstown" was the cry. Crowds came, property changed hands, old houses were razed, and by summer Millerstown was a modern oil town. It held on bravely until Bradford overwhelmed the southern region.

Toll Gate, Plank Road from Titusville to Pleasantville

Empire Transportation Company's Exhibit, Philadelphia Centennial, 1876

Only a few oil firms had exhibits at the Philadelphia Centennial Exposition in the summer of 1876. The Empire Transportation Company had the largest and most important—models of pipe lines, storage tanks, tank cars and terminal facilities

The history of the petroleum industry records many extraordinary and unprecedented events. One of these occurred at Titusville in the spring of 1877 just as Bradford and Bullion were beginning to attract attention. Early in April 1877, A. M. Heron commenced digging a hole in the ground about four feet square on the Giles farm a quarter of a mile east of Titusville's city limits and just below the plank road to Pleasantville. When down twenty feet on April 6th, Robert Powell, who was working at the bottom of the shaft, suddenly cried for help as the hole was rapidly filling with gas and oil. Powell was rescued and the oil continued to rise until it came within a few feet of the surface. Applying a common hand pump, the well yielded at the rate of thirty barrels a day. News of the strike spread through the town and within two hours the plank road was lined with vehicles. Thirty to fifty dollars bonus and one-half the oil were offered for lots ten feet square. The quick returns, the small outlay needed to secure oil, and the large profit caused a rash of excitement to break out which rivalled Pithole in its palmiest days. Operators from Bradford and the Lower Region flocked to Titusville with their pockets bulging with money to invest. Stages to and from Titusville kept busy carrying visitors estimated at 1,000 to 1,500 a day. A whole army of diggers was at work on Bonanza Flats, as the place was called. By August, fourteen shafts had been sunk, all within one-fourth of an acre. In the center stood an engine with boiler which furnished power for pumping these wells. A network of vibrating walking beams, joined together in every conceivable way by bolts and links or tied by ropes formed, when in motion, the most novel sight ever seen in the oil region. The movement resembled a huge pile of disabled grasshoppers writhing and kicking upon the ground. The place was, therefore, dubbed "Grasshopper City." Operations were confined to a few acres, the wells were quickly exhausted, and the excitement was all over by September.

"Grasshopper City," Titusville, 1877

Oilmen at Well, Bonanza Flats, Titusville, 1877

BRADFORD, 1876, above and 1880, below

In 1837 the United States Lanc Company, owners of a quarter of a million acres of land in McKean anc adjoining counties, sent Colonel Levit C. Little from New Hampshire to look after its interests. He located in Mc Kean County on Tuna Creek, eigh miles from the Pennsylvania-New York line. Other settlers came to the valley and Littleton was founded, which in 1858 adopted the name of Bradford A mile north was another village callec Tarport.

Located in Tuna Creek valley in the midst of a vast amphitheatre, formed by the surrounding hills from 300 to 70(feet in height, Bradford had a popula tion of about five or six hundred in 1875. Before the end of 1875 wild catters had penetrated the oil bearing sand around Bradford and the rush soon began. Train loads of oilmen from every part of the lower country

rowded the streets and overran the iotels, and Main Street blazed at iight like a full-fledged frontier town. Strains of noisy music floated from he saloons, dance halls were thronged, ind gambling dens ran without moles-ation.

Operators, landowners, and drillers nade the Bradford House their head-quarters. Hundreds of big contracts vere closed in the second story room where Lewis Emery, "Judge" Johnson, ind the advance guard assembled.

The population of Bradford in 1878 ipproximated 4,000. Over 1,000 letters were received daily at the post office; he money order business amounted to 32,000 a week. The telegraph company eceived and delivered as many as 500 nessages in a day. Railroad traffic was remendous; at times there were as nany as 250 loaded cars standing in he yards. Two locomotives were en-gaged night and day in switching.

Bradford House, Bradford, 1877

380

"Peg-Leg" Railroad and its End

With 7,000 producing wells in the district averaging 65,000 barrels a day Bradford was the oil metropolis of the world in 1880. Its population exceeded 11,000 and its post office was the third largest in Pennsylvania.

One of the unusual spectacles at Bradford was the Bradford & Foster Brook Railroad, better known as the "Peg-Leg." Built in 1877, the cars ran on an elevated track on a single rail. This rail was nailed to a wooden stringer which rested on top of piles. The cars, which set on the track like a saddle, ran on two double-flanged wheels, and were steadied by small wheels on either side, running on a heavy plank spiked and bolted to the piles. When completed in December 1877, the railroad ran day and night between Bradford and Tarport and later to Gilmor City.

In August 1878, as the "Peg-Leg" started to cross Tuna Creek, the sleeper on which the rail rested gave way, upsetting the two flatcars and the engine. Of the sixty passengers aboard, twenty were thrown into the water but no one was seriously injured. In February 1879, the railroad was sold at a Sheriff's sale.

Columbia Conduit Company Outwits the Railroad. In the spring of 1874 Dr. David Hostetter of Pittsburgh, widely known for "Hostetter's Stomach Bitters," began constructing a three inch pipe line from Millerstown to Fairview near Pittsburgh. The line crossed under the tracks of the Pennsylvania Railroad near Fairview, for which Hostetter claimed he had a right-of-way. Soon after operations commenced, the railroad sent a crew to wreck the pipe line. It was repaired only to be torn up again. Finally, J. G. Benton of Titusville suggested the solution. The pipe line crossed the railroad at a highway, so they hauled the oil across the tracks on the public highway in tank wagons, dumped it into the storage tanks, and then pumped it on to Pittsburgh.

BYRON D. BENSON

First President of the Tidewater
ipe Company, Limited, which was
rganized in Titusville on November
3, 1878, for the purpose of building
pipe line from the Bradford field to
Villiamsport, Pennsylvania. An ar-
angement was made whereby the
Reading Railroad would then haul
he crude oil in tank cars to Phila-
lelphia and New York.

DAVID K. MCKELVY

As the first counsel for the Tide-
ater, McKelvy piloted the company
rough all of its legal difficulties and,
collaboration with J. G. Benton,
esigned tongs, pipe jacks, jack boards,
vabs, and other equipment needed to
y the line. McKelvy succeeded Ben-
on as president in 1888.

Sledging the Pipe for "Benson's Folly"

Up until this time oil had never been pumped in a pipe larger than three inches nor more than thirty miles nor up and over any high elevation. The Tidewater line was to be a six inch line, 109 miles long, and oil was to be pumped over a mountain nearly 2,600 feet high. Skeptics called the new project "Benson's Folly" and made bets that the project would fail.

Much of the pipe had to be hauled from twenty to thirty miles over the mountains and frequently over unbroken roads in severe winter weather. The laying of the first pipe line across the Alleghenies in the dead of winter is one of the great feats in the history of the petroleum industry. On May 28, 1879, the line was completed; the pumps at Coryville were started, and the oil began to flow. After some initial delay, pumping was resumed and on June 4 the first oil reached its destination at Williamsport.

The first section of the Tidewater extended to Muncy Station and was completed in May 1879. The line was extended to Tamanend in the summer of 1882 and five years later to Bayonne, New Jersey.

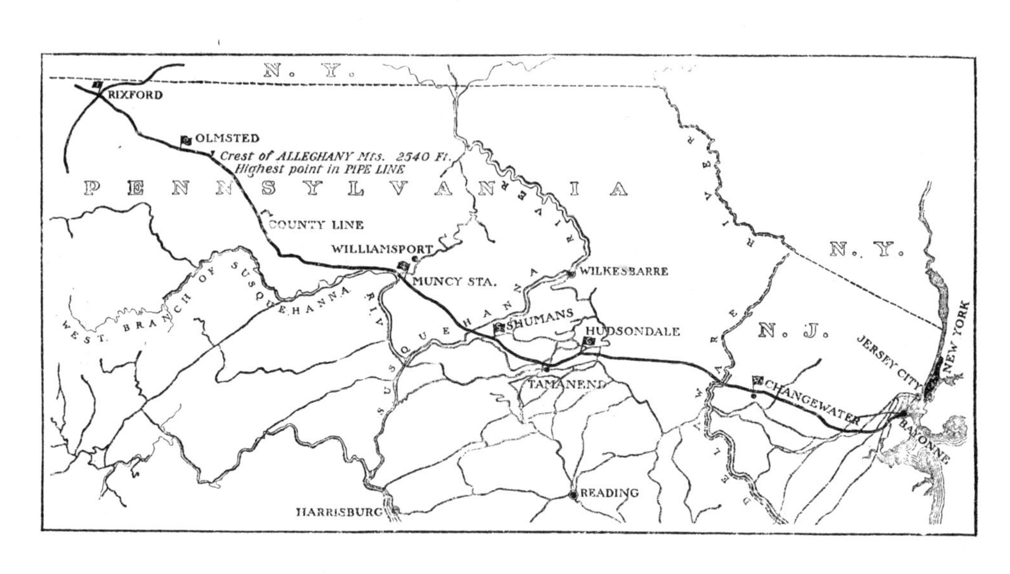

First Oil Exchange in Bradford, 1878. In March 1878, the Bradford Oil Exchange was organized with C. L. Wheeler as President. A new building was erected and dedicated in February 1879. It became the foremost oil exchange in the world. At the high water mark, transactions exceeded $1,000,000 a day.

Captain J. T. Jones Juvenile Band, Bradford. Although the town band practiced diligently, it lacked many instruments and uniforms. Captain J. T. Jones, the biggest individual producer in the Bradford field, furnished the money for these items. When the band made its next public appearance Jones' gift was well advertised by the inscription on the drum.

Lewis Emery, jr.

Roberts Torpedo Company Factory, Bradford

Stockholders and Directors of the Pure Oil Company, 1903-1904

In 1865 Lewis Emery, Jr., came to Pennsylvania and made his first venture in producing oil on Oil Creek at Pioneer. Bankrupt as a result of the Panic of 1873, Emery went to Bradford in 1875. He secured the Quintuple tract of 5,000 acres and drilled a test well three miles south of town. Its success confirmed his judgment of the territory and marked the beginning of the Quintuple development. It quickly placed Emery in the millionaire class. Emery built refineries, opened immense hardware stores, constructed pipe lines, and next to Captain J. T. Jones, was rated as the largest individual producer. In 1878 he was elected to the lower house of the state legislature and then served two terms in the Senate. Whether in or out of the legislature, Emery fought bitterly against the Standard Oil Company and championed the producers.

As a result of the oil excitement at Bradford many large oil companies were formed. One of these, whose history comes down to the presen[t] was the Pure Oil Company. A grou[p] of independent Bradford producer[s] including David Kirk, Lewis Emer[y] Jr., Michael Murphy, J. W. Lee an[d] others, who opposed the Standard O[il] Company, organized the Pure O[il] Company in 1895.

TARPORT AND TUNA VALLEY, 1880. Located about a mile and a half north of Bradford, the village of Tarport, when the oil rush came, rapidly grew and in 1878 it had the reputation of "the wickedest town of them all."

Hogan's House in Tarport

Killing Time in Tarport

There were many disreputable places of business in Tarport but Hogan's was the worst.

Richburg, New York, 1881

The Bradford field was extended to Richburg, Allegany County, New York, in 1881. On May 30 of that year, O. P. Taylor completed a well in a ravine close to the quiet village of Richburg; it started at thirty barrels and created much excitement. Samuel Boyle's 300 barrel well in July was overwhelming. Richburg and Bolivar, both old villages, quadrupled their size in three months. During the next eighteen months the limits of the field were defined, and a gradual decline began. The Allegany field was the northern limit of oil development in the United States.

Acme Oil Company Refinery, Olean, New York

After the destruction of its refinery at Titusville in the fire of 1880, the Acme never fully rebuilt the plant. Instead, it built a large refinery at Olean, close to the great Bradford field.

Map of the Western Pennsylvania Oil Region

Cherry Grove

When Captain Peter Grace and George Dimick began drilling a wildcat well in a dense forest on Lot 646, Cherry Grove township, Warren County, it resulted in one of the most exciting chapters in the history of the petroleum industry. The spot was nine miles from Clarendon and an equal distance from Sheffield. In March 1882, the report spread that the Grace & Dimick well on 646 had shut down, the derrick had been boarded up, and guards had been stationed around it to warn away visitors.

The well was made a "mystery," and it was the principal topic of conversation among oilmen. Drilling was resumed in May and on the 19th the oil trade was paralyzed by the report that it was flowing 1,000 barrels, then 2,500. It was as if an earthquake had hit the market. The excitement in the oil exchanges was indescribable. Over 4,500,000 barrels of oil were sold in one day on the exchanges in Titusville, Oil City, and Bradford. The Murphy, Cadwallader, Mehoopany and scores of other wells drilled at Cherry Grove ran the daily yield up to 40,000 barrels in August. Cherry Grove completely demoralized the market and drove the price down to 49½ cents, the lowest in years. By September the excitement was all over; dry holes and dwindling production brought about a collapse and Cherry Grove passed into history.

Famous "Mystery Well"—No. 646, Cherry Grove, Warren County, 1882

GARFIELD, CHERRY GROVE, 1882. With the completion of well No. 646, men came from everywhere. Two towns began to develop, Garfield at the end of the Clarendon plank road and Farnsworth near the end of the Sheffield plank road. At first, more than a mile apart, they grew until they met. By July 1 they had a population of 5,000 and in another month, 10,000 or more. Business was transacted under canvas tents, and at night men slept by the hundreds in the open field.

Pumping Station at Garfield

Famous Oil Scouts, Cherry Grove, 1882
ᵎft to right, seated, J. C. Tennent, Owen Evans, Dan
ᵋron, J. C. McMullin, and John Drake; standing, S. B.
Hughes, J. P. Cappeau, and Jule Rathburn

The professional oil scout first became prominent at
herry Grove. Prior to this time the newspapers of the
gion had been depended upon to furnish information
ncerning new wells and new fields but with the growth
wild speculation in oil as the Bradford field opened, the
arket was adversely or favorably affected by rumors
d "mystery" wells. In order to protect themselves the
ading interests in the oil trade employed trained men to
atch the wells and report daily on conditions. Oil scouts
owled at night and spied on "mystery" wells, watched
tels and livery stables to observe all who came or went,
tened in telegraph offices and ciphered messages being
nt or received, crept inside the guard lines to gauge a well,
d resorted to all sorts of measures to secure accurate in-
rmation for their employers.

Mystery wells were heavily guarded day and night and scouts had to keep concealed at a safe distance. If a twig cracked, bullets whizzed in the direction of the noise.

Thorn Creek

After Bradford reached its peak i production, Thorn Creek, about si miles southwest of Butler, became th next prolific source of crude oil. I 1884 the Phillips brothers starte drilling on the Barlett farm on Thor Creek. They hit the sand on Augus 29 and the well flowed 500 barrels Thousands of people came to see th gusher each day. Drilling ten fee deeper, the well flowed 4,200 barrel on September 15. The market wa demoralized and the price of o dropped eight cents a barrel. The rus to Thorn Creek was under way.

On October 11 drilling at the Chri tie well, 360 feet west of the Phillip was near the critical point. Exciteme was at fever heat among those wh watched the operations. In the afte noon someone reported that the dri was 27 feet through the sand and n oil. Many of the scouts wired the employers that the well was dr Within a short time, the well bega to gas and then a cloud of oil e veloped the derrick. The Christie we was not dry. It was the biggest gush the oil country had ever known! Th first day it produced over 5,000 ba rels, 7,000 for several days after bein torpedoed, and for a month it poure out a sea of oil. One hundred thousan dollars was refused for the well. Oth good wells were completed withi the next ten days and everyone wo dered what would happen next.

Phillips Well, Thorn Creek, Butler County, 1884

Christie Well, Thorn Creek, 1884

Oil Scouts at Phillips Well, Thorn Creek

A telegraph office was rigged up i an abandoned carriage near th Phillips well; sharp-eyed oil scou thronged about it day and night.

Colonel S. P. Armstrong had lease a portion of the Marshall farm c Thorn Creek and, in October, wa drilling his second well about 400 fe south of the Phillips. The drill ha gone through the sand and the drillin contractor pronounced the well a fai ure, but the owner did not lose hop He decided to use a torpedo. At noo on October 27 sixty quarts of nitr glycerine were lowered into the hol There was a low rumble but the we did not flow. The scouts ran for th telegraph office to wire that the we was dry. The rumbling in the well i creased and in a moment oil cam out with a mighty rush. A giant strea spouted sixty feet above the derric and saturated everything within radius of 500 feet. During the fir twenty-four hours the Armstron flowed 8,800 barrels! It dropped 6,000 by November 1, to 600 b December 1, and stopped complete the next day.

Armstrong Well No. 2, Thorn Creek, 1884

CHAPTER X

GREAT OIL FIRES

THERE were many fires in the oil region like the burning of the Drake well in October 1859, the Rouse fire of April 1861, and the Grant well fire at Pithole in the summer of 1865. Two were memorable disasters. The first occurred at Titusville on Friday morning, June 11, 1880. During a severe thunderstorm a flash of lightning struck Tank 3 of the Titusville and Tidioute Pipe Line, situated on the peak of the south side hill, west of the foot of Perry Street. Instantly a dense cloud of smoke and flame shot skyward and 20,000 barrels of oil were afire. Two other storage tanks ignited from the intense heat and exploded with a tremendous report. Flaming torrents of oil ran down the hill to the Acme Oil Company plant No. 1 and, as tank after tank exploded hurling mountains of smoke and flame upward, Titusville appeared to the awe-stricken citizens a doomed city. For three days the fire raged with undiminished fury but the city was miraculously spared. Not a single life was lost in the disaster.

The worst disaster took place on Saturday, June 4, 1892, and involved both Titusville and Oil City. For several days preceding, there had been heavy rains in Oil Creek valley. On Saturday morning Oil Creek had risen to the top of its banks; by midnight, water had flooded many of the streets in Titusville and it was without water, gas, and electricity. About midnight a large dam at Spartansburg, a few miles above Titusville, gave way. With a roar like thunder, a mass of water struck Titusville. It came with such suddenness that people did not have time to reach higher ground. By 2 a.m. the whole southern portion of town as far north as the main business section was under water. Homes floated away from their foundations and hundreds of people sought safety on roofs, telephone poles, trees, and drifting timbers. To add to the horror, fire broke out in the Schwartz refinery about 3 a.m. Tanks and stills were blown into fragments; streams of burning oil on the surface of the water carried the flames up and down the Creek; and there were terrifying explosions at neighboring refineries. Oil Creek was a sheet of fire. Thousands of panic-stricken people fled to the hills as the fire spread. The fire raged all day Sunday but was brought under control on Monday.

Hundreds of Oil City citizens lined the banks of Oil Creek on Sunday morning watching the rapidly rising water. A bluish haze hung over the raging waters—benzine gas. Suddenly there was a blinding flash and instantly the Creek from Rouseville to Oil City was a mass of roaring flames and billowing clouds of smoke rolled high over Oil City. Wild with terror, people wept and fainted; thousands ran to the hills for safety.

FIRE AT THE GRANT WELL, 1865

General View, Acme Oil Company Refinery Ruins, 1880

The Acme Oil Company suffered the greatest loss in the fire of 1880; its property damage amounted to more than $275,000. Active operations of the company were never restored, except in a limited measure. The company built an extensive plant at Olean, New York, and turned its back upon Titusville forever.

Acme Refinery Ruins, 1880

Acme Refinery Ruins, 1880

Acme Refinery Ruins, South Perry Bridge, 1880

TITUSVILLE FIRE AND FLOOD, 1892

TITUSVILLE FIRE AND FLOOD, 1892

Fire and Flood Ruins, Titusville, 1892

Fire and Flood Ruins, Titusville, 1892

When the forces of destruction had spent themselves, over seventy-five lives had been lost, hundreds made homeless, and the property damage amounted to more than $1,000,000 in Titusville; in Oil City, over fifty people lost their lives, eight hundred were homeless, and the property damage approximated that of Titusville.

Fire and Flood Refugees, Titusville, 1892

Railroad Tracks and Flood Ruins, Titusville, 1892

Seneca Street, Oil City, before Fire, 1892

OIL CITY FIRE AND FLOOD, 1892

Street Scene, Oil City, before Fire, 1892

Centre Street Bridge, Oil City, 1892

Centre Street, Oil City, 1892

North Seneca Street, Oil City, 1892

Titusville after the Fire

Railroad Bridge, South Martin Street, Titusville, 1892

Oil City after the Fire

CHAPTER XI

HONORING DRAKE AND THE BIRTH OF THE PETROLEUM INDUSTRY

FROM the time Drake completed his famous oil well until 1934, various movements originated in the northwestern Pennsylvania oil region to honor Colonel Drake and the birth of the petroleum industry. The first movement realized its goal on October 4, 1901, when a magnificent monument to the memory of Drake, the generous gift of Henry H. Rogers, was unveiled and dedicated in Woodlawn Cemetery, Titusville. A short time thereafter the body of Colonel Drake was exhumed at Bethlehem, Pennsylvania, and removed to a position in front of the Rogers monument.

The second movement, one to mark the site of Drake's well, was inaugurated by Canadohta Chapter, Daughters of the American Revolution, Titusville. About 1908 Mrs. David Emery donated to the Chapter one acre of land upon which the Drake well had been drilled. Upon this spot the organization placed a native boulder weighing about thirty tons, and on August 27, 1914, a large bronze tablet was set in the stone with a replica in bas-relief of the Drake well and an inscription dedicating the boulder.

Simultaneous with the efforts of the Daughters of the American Revolution was the movement to memorialize this epoch-making event by something more than an inanimate marker. Furthermore, there was a conviction that it should be done by oilmen. This led to the concept of a museum and library relating to petroleum history. The central figure at the outset was E. C. Bell of Titusville, who had spent fifty years collecting all sorts of historical material relating to the early history of the industry. Through his efforts a small museum was built about 1915 near his home just west of Titusville. However, after his death in 1923, the museum was discontinued and the materials placed in storage.

The idea of a museum and library did not die. A group of public spirited and historically minded citizens in the oil region decided that a museum and library should be erected at the site of the Drake well and that it should become a national depository for all kinds of historical relics and records pertaining to the early history of the petroleum industry. As a result, the American Petroleum Institute raised $60,000 in 1931 and executed the idea. In 1934, when all the work had been completed, the property was given to the Commonwealth of Pennsylvania. Today it is a state park with some 229 acres under the administrative jurisdiction of the Pennsylvania Historical and Museum Commission.

Mather at the Site of Drake's Well, August 16, 1896

Unveiling the Drake Monument, Woodlawn Cemetery, Titusville, October 4, 1901

Erected by Henry H. Rogers, the Drake monument cost $100,000. Charles Brigham designed the memorial and Charles Henry Niehaus made the central bronze figure.

Breaking Ground for Drake Museum, October 3, 1911. R. D. Fletcher, long a merchant in Titusville and an intimate friend of Drake, is shown holding the shovel.

Drake Museum, West Central Avenue, about 1915

E. C. Bell came to Titusville about ten years after the drilling of the Drake well. During the next fifty years he collected anything and everything pertaining to the early history of petroleum—a collection unequalled by any in the United States. This led to the idea of a museum and library as a memorial for Drake and the Drake well. The leading spirit in this movement was Bell.

With limited funds and on land generously donated, a small brick building was soon completed near Bell's home, just west of Titusville and on the bank of Oil Creek. At an age of more than sixty, Bell did most of the construction work himself. When it was completed, he deposited his collection of historical materials and relics in the museum.

E. C. BELL

Bell and Friends of the Museum

Bronze Plaque, Drake Well Monument. In order to preserve the site of the Drake well, this native limestone boulder, weighing about thirty tons, was placed here in 1914 by Canadohta Chapter, Daughters of the American Revolution.

Bell's idea of a museum and library did not die. In 1931 the American Petroleum Institute agreed to raise funds to construct a dike to keep the site of the Drake well from being flooded, clear the ground, excavate and drain the area, construct a caretaker's house, and establish a museum and library, provided the Commonwealth of Pennsylvania would accept the property as an historical park and appropriate an annual sum for its proper maintenance and development.

Displays featuring oil lamps, drilling tools, and all kinds of historical documents and relics are to be seen in the picture below. The library contains rare books, newspapers, oil company records, pamphlets, photographs, and documents on oil history. Altogether there are 3,274 Mather negatives in the Drake Museum.

Drake Well Memorial Park near Titusville

The Drake Museum is a treasure house of all sorts of historical materials and records pertaining to the history of the petroleum industry. It is the largest depository in the country devoted exclusively to petroleum history.

When all improvements had been completed in the summer of 1934, the Commonwealth formally accepted the gift. To the extreme right is the Drake Well Monument; in the center is the caretaker's house; and to the left is the Museum.

Replica of the Drake Well, Drake Well Memorial Park, 1945

Many of the Drake Well Memorial Park visitors were disappointed not to see something of the old Drake well. The Drake Well Monument was therefore moved to the west side of the central oval and an exact reproduction of Drake's derrick and engine-house was erected in 1945 on the spot where Drake drilled.

LIST OF PHOTOGRAPHS

CHAPTER I

Earliest Record of Petroleum in Pennsylvania. From Lewis Evans' *Map of the Middle British Colonies in America* published in 1755
First Petroleum Shipped to Pittsburgh. From J. J. McLaurin, *Sketches in Crude-Oil*, p.16
Early Quotation on Oil, 1797. From the Day Book of General William Wilson in the John Reynolds Collection, Meadville, Pennsylvania
Collecting Oil on Oil Creek, 1810. From a print in the Western Pennsylvania Historical Society
Samuel M. Kier. Drake Museum Collection No. 1147. See also No. 1148
Kier's Advertisement. From J. T. Henry, *The Early and Later History of Petroleum*, p.83
Kier's Petroleum, or Rock Oil. Drake Museum Collection No. 1662
Kier's Circular. Drake Museum Collection No. 1663
Francis Beattie Brewer. From J. T. Henry, *The Early and Later History of Petroleum*, facing p.393
George H. Bissell. From J. T. Henry, *The Early and Later History of Petroleum*, facing p.346
Silliman's Report. From an original copy in the Drake Museum
Benjamin Silliman, Jr. From Yale University Library
James M. Townsend. Drake Museum Collection No. 1152
Seneca Oil Company Stock Certificate. From original in the Drake Museum
Edwin L. Drake. Mather Collection No. 1146
William A. Smith. Mather Collection No. 1150
The Drake Well in 1861. Mather Collection No. 4
Type of Engine and Boiler Used by Drake. From the *Oil and Gas Journal*, August 23, 1934, p.24
Drake's Drilling Tools. From an original photograph by Mather in the Drake Museum
The Drake Well in 1864. Mather Collection No. 456

CHAPTER II

Lower Oil Creek and Vicinity. From Paul H. Giddens, *The Birth of the Oil Industry*, The Macmillan Company, 1938, p.79
Locating an Oil Well—Witch Hazel Method. Mather Collection No. 609
Kicking Down a Well. From *The Oil and Gas Man's Magazine*, VII, April 1912, p.317
Kicking Down a Well. From *McClure's Magazine*, February 1903, p.398
Sketch of a Well Being Drilled. From S. J. M. Eaton, *Petroleum: A History of the Oil Region of Venango County, Pennsylvania*, p.112
Steam Engine for Drilling. From an advertisement in the *Titusville Morning Herald*, September 19, 1865
Inside View of a Derrick. From *Frank Leslie's Illustrated Newspaper*, January 21, 1865
The Empire Well, 1863, Funk Farm. Mather Collection No. 1480. See also No. 808
Phillips and Woodford Wells, Tarr Farm. From a photograph made by Mather in December 1861. In J. T. Henry, *The Early and Later History of Petroleum*, Frontispiece
Underground Storage Tanks, Tarr Farm. Mather Collection No. 1587
The Sherman Well, 1863, Foster Farm. Mather Collection No. 753
The Noble and Delamater Well, 1863, Farrel Farm. From Charles A. Whiteshot, *The Oil Well Driller*, p.57
Tarr Farm, 1864. Mather Collection No. 1858
A Women's Party, Oil Creek, 1864. Mather Collection No. 768
A Barrel Yard, Shaffer Farm. Original print in Drake Museum, P.883
Loading Platform, Shaffer Farm. Original print in Drake Museum, P.858
"Fishing" for Lost Drilling Tools, Oil Creek. Drake Museum Collection, P.899
Oil Dippers at Miller Farm, 1863. Mather Collection No. 1684
"Oil on the Brain." Original copy in the Drake Museum
Barrel Factory, Oil Creek. Original print in Drake Museum
Cow Run Wells, 1864. Mather Collection No. 1550
Bull Run Wells, 1863. Mather Collection No. 1802
Magill Wells, Foster Farm, 1864-1866. Mather Collection No. 1796
Pioneer Run Wells, 1865. Mather Collection No. 1820
An Oilman's Home, Pioneer Run. Mather Collection No. 551
Flatboat with Cargo of Oil for Pittsburgh, Pa. From *Frank Leslie's Illustrated Newspaper*, January 21, 1865
Great Western Run Wells, 1864-1868. Mather Collection No. 1818
Monitor Refinery, G. W. McClintock Farm, 1864. Mather Collection No. 751
Boyd, Benninghoff, and Stevenson Farms, Petroleum Centre, 1864-1868. Mather Collection No. 1800
Benninghoff Farm, Benninghoff Run, 1865. Mather Collection Nos. 1799, 1805
Wooden Storage Tanks, Benninghoff Run, 1865. Mather Collection No. 555
Horse-drawn Flatcar, Benninghoff Run. Original print by Mather in Drake Museum
The Benninghoff Robbery. From the *Titusville Morning Herald*, January 17, 1868
King of the Hills Well, Stevenson Farm, 1867. Mather Collection No. 678
James S. McCray Farm and the Egbert Tract, Petroleum Centre, 1873. Mather Collection No. 1780
Columbia Oil Company Farm, 1867. Mather Collection No. 1830
Farm Manager's Office, Columbia Oil Company, 1864. Mather Collection No. 606
Machine Shops, Columbia Oil Company, 1868. Mather Collection No. 607
Workmen on Oil Creek, 1865. Mather Collection No. 756
Columbia Cornet Band, 1867. Mather Collection No. 1829
Tarr Farm, 1865-1868. Mather Collection No. 1798
James Tarr's Residence, Tarr Farm, 1866. Mather Collection No. 1109
Boiler and Engine "For Sale Cheap," Tarr Farm. Mather Collection No. 680
Group at Tarr Farm, Mather Collection No. 749
Blood Farm House, 1864. Mather Collection No. 638
Blood Farm Ferry, 1868. Mather Collection No. 765
Twin Wells, Blood Farm, 1863. Mather Collection No. 1565
Niagara & Pierson Farms, Cherrytree Run, 1868. Mather Collection No. 1826
Musicians and Oilmen, Hess Farm Wells, Cherrytree Run. Mather Collection No. 1486
Rynd Farm Wells, 1865. Mather Collection No. 1609
Allemagoozelum Well, Cherrytree Run, 1869. Mather Collection No. 1496
Hess Farm Wells, Cherrytree Run, 1869. Mather Collection No. 1482
"Coal Oil Johnny's" Farm, 1864. Mather Collection No. 556. See also No. 1572
"Coal Oil Johnny." Mather Collection No. 1151
"Coal Oil Johnny's" Bankruptcy. From the *Titusville Morning Herald*, February 15, 1868

Dangerous Character of Petroleum. Original copy in the Drake Museum
"The Amateur Millionaire, or What Came of an Oil Strike." Original copy in the Drake Museum
Program of "Struck Oil." Original in the Drake Museum
"Struck Oil." *Titusville Morning Herald*, November 3, 1877
Program of "Struck Oil." Original in the Drake Museum
Cherry Run, 1867. Mather Collection No. 1827
Cornplanter Run Wells, 1864. Mather Collection No. 1568
The Petroleum Producers' Association. Original certificate in the Drake Museum

CHAPTER III

The Oil Region 1859-1870. From Paul H. Giddens, *The Birth of the Oil Industry*, The Macmillan Company, 1938, p.50
Guide to the Oil Regions, 1865. Original in the Drake Museum
Corry, Pennsylvania, 1863. Mather Collection No. 663
"Rush to the Cars at Corry." From *Frank Leslie's Illustrated Newspaper*, January 7, 1865
Downer Oil Works, Corry. Mather Collection No. 665
General View of Titusville, 1864. Mather Collection No. 1710. See also No. 1711
North Side, Spring Street, Titusville, 1865. Mather Collection No. 589
South Side, Spring Street, Titusville, 1865. Mather Collection No. 717
Spring Street, Looking West, Titusville, 1865. Mather Collection No. 722
South Washington Street, Titusville, 1865. Mather Collection No. 700
Hinkley Refinery, 1863. Mather Collection No. 706
Crittenden House, Titusville. Mather Collection No. 728
First Daily Newspaper in the Oil Region. Original print by Mather in the Drake Museum
Sleeping in Chairs at Hotel, Titusville, 1865. From *Frank Leslie's Illustrated Newspaper*, January 7, 1865
Corinthian Hall, Titusville. Mather Collection No. 718
Brown's Band, Titusville, 1868. Mather Collection No. 738
Mather's Boat. Mather Collection No. 564. See also No. 684
Mather's Wagon. Mather Collection No. 291
Mather's Advertisement. *Titusville Morning Herald*, July 28, 1866
Interior View, Mather's Photograph Gallery, 1869. Mather Collection No. 983
Alvord House, Miller Farm. Mather Collection No. 778
Indian Rock Oil Company Office, Miller Farm. Mather Collection No. 1614
McElheny Oil Company and Dalzell Petroleum Company Office, Petroleum Centre, 1863-1864. Mather Collection No. 1582
Funkville, 1867. Mather Collection No. 1822. See also No. 1821
Pioneer and Railroad Bridge, 1865. Mather Collection No. 1815
Oilmen at the Erie Hotel in Pioneer, 1868. Mather Collection No. 610
Boughton Cooperage Shop, Pioneer, 1864. Mather Collection No. 782
Brown's Hotel, Pioneer. Original print in the Drake Museum
McKinney Oil Office, Pioneer, 1864. Mather Collection No. 613
Office of the Union Oil Company and the Hoskins Oil Company, Pioneer, 1867. Mather Collection No. 784
D. C. Clark's Store, Pioneer, 1864. Mather Collection No. 786
Boarding House at Pioneer. Original print in the Drake Museum
Petroleum Centre, 1864. Mather Collection No. 1776
Central Petroleum Company Office, Petroleum Centre, 1865. Mather Collection No. 804
Refinery at Petroleum Centre, 1864. Mather Collection No. 626
Central House, Petroleum Centre. Mather Collection Nos. 1592, 1779
J. J. Stoltz Boot & Shoe Shop, Petroleum Centre, 1868. Mather Collection No. 621
George H. Bissell & Company Bank, Petroleum Centre, 1868. Mather Collection No. 806
Washington Street, Petroleum Centre, 1868. Mather Collection No. 1599
Petroleum Centre Reunion Picnic, 1890. Mather Collection Nos. 2123, 2124
McClintockville, 1861. From A. C. Crum, *Romance of American Petroleum and Gas*, I, p.80
Street Scene, Tarr Farm, 1865. Mather Collection No. 270
Street Scene, Rouseville, about 1867. Mather Collection No. 279
Rouseville, 1867. Mather Collection No. 1849
Main Street, Oil City, 1861. From A. C. Crum, *Romance of American Petroleum and Gas*, I, p.77
Cottage Hill, Oil City, 1863. Mather Collection No. 1504
Long Bridge over the Allegheny River, Oil City, 1864. Mather Collection Nos. 1099, 1503
Hogback Hill, Oil City, 1863-1864. Mather Collection No. 1500
The Oil City Derrick. Original in the Drake Museum
Union Station, Oil City. Original print in the Drake Museum
Agitator, Brundred Refinery, Oil City. Mather Collection No. 469
Commercial Hotel, Oil City. Mather Collection No. 2138
T. M. George's Keystone House, Franklin. Mather Collection No. 667

CHAPTER IV

Teamster on Plank Road. Mather Collection No. 268
Teamsters Fording Oil Creek. Mather Collection No. 558
Bad Roads, Sherman Well, Oil Creek, 1863. Mather Collection No. 1574
Loading Oil at Funkville. Mather Collection No. 552
Filling Barrels with Oil. From *Frank Leslie's Illustrated Newspaper*, January 21, 1865
Filling Flatboats with Oil. From *Frank Leslie's Illustrated Newspaper*, January 21, 1865
Pond Freshet Boat Hits Bridge Pier, Oil City, May 1864. Original print in the Drake Museum
Pond Freshet Disaster, Oil City, May 1864. Original print in the Drake Museum
Pond Freshet Disaster, Oil City, May 1864. Original print in the Drake Museum
The Oil Fleet at Oil City. Mather Collection No. 1560
Transporting Barrels on Oil Creek. From *The London Illustrated News*, February 27, 1875
Getting Flatboats Up Stream. From *Frank Leslie's Illustrated Newspaper*, January 28, 1865
Packet Express Boat, 1863. Mather Collection No. 1707
Passenger Packet Going Down Oil Creek. From *Leslie's Illustrated Newspaper*, January 21, 1865
Bad Roads. From *Frank Leslie's Illustrated Newspaper*, February 11, 1865
Teaming on Oil Creek. From *Frank Leslie's Illustrated Newspaper*, January 28, 1865
Map of the Oil Region, 1865. From *Derrick and Drill*
Locomotive "Oil City," Oil Creek Railroad. Mather Collection No. 793
First Oil Cars. Mather Collection No. 1100
Oil Creek Railroad Locomotive at Petroleum Centre. Mather Collection No. 302
Oil Creek Railroad Locomotive on the Bridge, Pioneer. Mather Collection No. 792

CHAPTER V

United States, or Frazier, Well. Mather Collection No. 1770
Astor House. Mather Collection No. 1763
United States Petroleum Company Office, 1865. Mather Collection No. 1773
John Wilkes Booth. From an original print in the Drake Museum
Homestead Well. Mather Collection No. 1524
Pithole and Balltown, 1865. Mather Collection No. 1767
Ellsworth Oil Well Association Certificate. From the original certificate in the Drake Museum
Grant Well Office. Mather Collection No. 1094
First Street. Mather Collection No. 640
Pithole, 1865. Mather Collection No. 1520

Pithole, 1865. Mather Collection No. 1765
Eureka Well and Office. Mather Collection No. 639
Well at Pithole. Mather Collection No. 1087
Charles Pratt, Captain of the Pithole Swordsman's Club. Mather Collection No. 163
Bonta House. From an original print in the Drake Museum
Ben Hogan, "The Wickedest Man in the World." From *The Flaming Ben Hogan*, front cover. Courtesy of Nat Fleischer
"French Kate." *The Flaming Ben Hogan*, p.5
The Life and Adventures of Ben Hogan, by George F. Trainer. Courtesy of R. W. Tillotson
Pithole Pioneers Reunion. Mather Collection No. 1917
Dining Hall at Pithole. Mather Collection No. 1088
Holmden Street. Mather Collection No. 641
Corner of Holmden and First Streets, August 1895. Mather Collection No. 1523
Pithole, 1934. Drake Museum Collection No. 1518

CHAPTER VI

Hauling Oil at Titusville before 1866. From *Frank Leslie's Illustrated Newspaper*, January 7, 1865
First Tank Cars—Densmore Type. Mather Collection No. 570
Empire Transportation Company Advertisement. From the *Titusville Morning Herald*, September 3, 1877
Iron Boiler Tank Car—1880. Mather Collection No. 1869
Samuel Van Syckel. From an original print in the Drake Museum
Reed and Cogswell Steam Pump. From A. C. Crum, *Romance of American Petroleum and Gas*, I, p.95
Pipe Line Dump—Pithole. From A. W. Smiley, *A Few Scraps, Oily and Otherwise*, p.125
Miller Farm, Oil Creek, 1865-1868. Mather Collection No. 1781
Henry Harley. From J. T. Henry, *Early and Later History of Petroleum*, facing p.526
W. H. Abbott. From J. T. Henry, *Early and Later History of Petroleum*, facing p.360
Abbott & Harley Pipe Line Terminal, Shaffer Farm. Mather Collection No. 1834
Allegheny Transportation Company Office, Miller Farm, 1868. Mather Collection No. 1478
Pennsylvania Transportation Company Office, Miller Farm. Mather Collection No. 1782
Oil Creek Railroad Depot, Titusville, 1865. From *Frank Leslie's Illustrated Newspaper*, January 21, 1865
The Reed Well, Smith Farm, 1863. Mather Collection No. 2161
Colonel E. A. L. Roberts. From J. T. Henry, *Early and Later History of Petroleum*, facing p.540
The Roberts Torpedo. From J. T. Henry, *Early and Later History of Petroleum*, facing p.251
First Roberts Torpedo Factory, Titusville. From the collection of the late E. T. Roberts
W. B. Roberts. From J. T. Henry, *Early and Later History of Petroleum*, facing p.417
Agreement to Organize the Roberts Torpedo Company. From the collection of the late E. T. Roberts
Roberts Torpedo Advertisement, *Titusville Morning Herald*, September 3, 1877
Delivering a Shot to a Well. From the collection of the late E. T. Roberts
Pouring the Shot. From the collection of the late E. T. Roberts
Cupler & Company's Wagon. Mather Collection No. 1670
Nitroglycerine Kills Cupler. Mather Collection No. 1444
A Torpedoed Well. Mather Collection No. 1647

CHAPTER VII

The Economite Wells, Tidioute, 1864. Mather Collection No. 1054
Tidioute, 1864. Mather Collection No. 1840
Train and Muddy Road, Tidioute. Mather Collection No. 655
Empire Line Office, Tidioute, 1867. Mather Collection No. 656
Oil Barges at Tidioute. Mather Collection No. 1055
Barrel Yard and Exchange Hotel, Tidioute. Mather Collection No. 817
Triumph Hill, East Side, near Tidioute. Mather Collection No. 1844
Wells on Dingley Run near Tidioute. Mather Collection No. 661
Triumph Oil Company Office, Triumph. Mather Collection No. 819
King Bee Well, Triumph Hill. Mather Collection No. 1839
Dr. G. S. Shamburg's Home, Shamburg. Mather Collection No. 1530
General View of Wells, Shamburg, 1868-1870. Mather Collection No. 1533
H. Spears Well, Shamburg Petroleum Company, 1868. Mather Collection No. 1794
The Fee Well, Atkinson Farm, Shamburg, 1868. Mather Collection No. 1541
Lady Stewart Well, Shamburg. Mather Collection No. 1548
Tallman Farm Office, Shamburg, 1868. Mather Collection No. 327
Beardsley House, Shamburg. Mather Collection No. 1529
Lyman Stewart's Home, Shamburg. Mather Collection No. 324
Main Street, Shamburg, 1868. Mather Collection Nos. 323, 1528
Post Office, Shamburg. Mather Collection No. 320
Hardware Store, Shamburg. Mather Collection No. 321
Red Hot near Shamburg. Mather Collection No. 1795
Stores at Red Hot. Mather Collection No. 318
Travelers Rest, Red Hot. Mather Collection No. 316
Harmonial Well, Pleasantville. Mather Collection No. 1512
General View, William Porter Farm, Pleasantville. Mather Collection No. 1509
Wells on the Armstrong Farm, Pleasantville. Mather Collection No. 1513
Oilmen and Drilling Tools, Pleasantville. Mather Collection No. 346
The "Corners," Pleasantville. Mather Collection No. 340
Chase House, Pleasantville. Mather Collection No. 1510
New York & Providence Petroleum Company Office, Pleasantville. Mather Collection No. 1514
Brown Brothers Store, Pleasantville. Mather Collection No. 688
Ned Pitcher Well, Pleasantville. Mather Collection No. 1515
First Large Pump Station near Pleasantville. From *The Oil and Gas Man's Magazine*, 1907, p.80. See also Mather Collection No. 1837

CHAPTER VIII

Titusville from South Hill, 1873. Mather Collection No. 1714
The Mansion House about 1874. Mather Collection No. 1700
West Spring Street and the Parshall Hotel about 1874. Mather Collection No. 1701
Granger & Company Store, Franklin and Central Avenues, 1868. Mather Collection No. 713
E. K. Thompson's Drug & Chemical Depot, 1867. Mather Collection No. 1013
Banquet for Henry Harley, Parshall Hotel. Mather Collection No. 595
John J. Carter in 1880, Founder of the Carter Oil Company. Mather Collection No. 1875
Stettheimer's Clothing Store. Mather Collection No. 1204
Carter's Clothing Store about 1869. Mather Collection No. 267
John J. Carter's Advertisement, *Titusville Morning Herald*, November 23, 1870
J. N. Pew's Advertisement, *Titusville Morning Herald*, June 3, 1872
Eames Petroleum Iron Works, Titusville, 1879. From *Frank Leslie's Illustrated Newspaper*, January 24, 1880, p.392
Excursion Train to Oil Creek Lake. Mather Collection No. 742
Acme Refinery No. 3. Mather Collection No. 1752
Acme Refinery Workmen. Mather Collection No. 1709
Titusville, 1884. Mather Collection No. 1717
Mather's Gallery. Mather Collection No. 1990
Barnsdall Oil Company Office. Mather Collection No. 1548
Heisman's Cooper Shop. Mather Collection No. 1183
Henderson's Oil Brokerage Office and Oilmen. Mather Collection No. 970

Titusville Oil Exchange. Mather Collection No. 974
The New Titusville Oil Exchange, West Spring Street. Mather Collection Nos. 976, 954
Oilmen in Front of the Oil Exchange Building. Mather Collection No. 971
St. James Memorial Church, and Appleton's Collegiate Institute. Mather Collection No. 729
F. S. Tarbell's Residence, East Main Street. Mather Collection No. 1256
Drake Street School. Mather Collection No. 1730
Eighth Grade of the Drake Street School, 1881. Courtesy of E. A. Schaaf
Celebrating the Centennial Exposition, Emerson's Residence. Mather Collection No. 1306
Mrs. George R. Harley and Mrs. J. H. Cogswell. Mather Collection No. 1906
D. H. Cady's Residence. Mather Collection No. 732
Mrs. John J. Carter. Mather Collection No. 1898
Charles Gibbs Carter as a Young Man. From an original print, courtesy of Mrs. Samuel Grumbine
Menu, Reception Supper, Sterrett-Farrel Wedding. From the original in the Drake Museum
J. C. McKinney's Residence. Mather Collection No. 1304

CHAPTER IX

James B. Kerr Farm, Church Run, near Titusville, 1870. Mather Collection No. 1783
John D. Rockefeller in 1872. From *McClure's Magazine*, December 1902, p.119
Peter H. Watson. From *McClure's Magazine*, January 1903, p.255
John D. Archbold in 1872. From *McClure's Magazine*, January 1903, p.256
An Oil Riot, 1872. From *McClure's Magazine*, January 1903, p.258
C. D. Angell. Mather Collection No. 359
Mutual Pipe Line Office, Foxburg. From A. W. Smiley, *A Few Scraps, Oily and Otherwise*, p.159
Foxburg in 1872. From A. W. Smiley, *A Few Scraps, Oily and Otherwise*, p.194
Pittsburgh Oil Brokerage, 1869. From *The Oil and Gas Journal*, October 4, 1923, p.68
Parker's Landing, 1874. Mather Collection No. 1705
Brokers and Oilmen, Parker's Landing. Mather Collection No. 1761
Hogan. From *The Life and Adventures of Ben Hogan*
Lizzie Toppling. From *The Life and Adventures of Ben Hogan*
Ben Hogan's Floating Palace, Parker's Landing. From *The Life and Adventures of Ben Hogan*, p.112
"French Kate." From *The Life and Adventures of Ben Hogan*
On Board Hogan's Floating Palace. From *The Life and Adventures of Ben Hogan*, p.118
Petrolia, Butler County. Mather Collection No. 1624
Fairview Pipe Line Company Office, Petrolia. From *The Oil and Gas Man's Magazine*, VI, May 1911, p.135
Main Office, United Pipe Lines, Petrolia. *The Oil and Gas Man's Magazine*, 1908, p.61
Karns City, Butler County. Mather Collection No. 1623
Millerstown, Butler County. Mather Collection No. 1629
Toll Gate, Plank Road from Titusville to Pleasantville. Mather Collection Nos. 1006, 1007
Empire Transportation Company's Exhibit, Philadelphia Centennial, 1876. From the *Historical Register of the Centennial Exposition*, p.307
"Grasshopper City," Titusville, 1877. Mather Collection Nos. 1020, 1741
Oilmen at Well, Bonanza Flats, Titusville, 1877. Mather Collection No. 1563
Bradford, 1876. From M. B. Rowe, *Captain Jones*, facing p.62
Bradford, 1880. Mather Collection No. 852
Bradford House, Bradford, 1877. From *Illustrated History of Bradford*, edited by V. A. Hatch, p.14
Bradford, 1880. Mather Collection No. 1851
"Peg-Leg" Railroad, Bradford.
Wreck of the "Peg-Leg," 1878. From *The Oil and Gas Man's Magazine*, VIII, April 1913, p.762
Columbia Conduit Company Outwits the Railroad. From an original print in the Drake Museum
Byron D. Benson. From A. C. Crum, *Romance of American Petroleum and Gas*, I.
Pipe for "Benson's Folly." From the original painting. Courtesy of the Tide Water Associated Oil Company
David K. McKelvy. From *McClure's Magazine*, January 1904, p.295
Route of the Tidewater Pipe Line. From *McClure's Magazine*, January 1904, p.303
First Oil Exchange in Bradford, 1878. From *Illustrated History of Bradford* edited by V. A. Hatch, p.31
Captain J. T. Jones Juvenile Band, Bradford. From M. B. Rowe, *Captain Jones*, facing p.112
Lewis Emery, Jr. Mather Collection No. 362
Roberts Torpedo Company Factory, Bradford. From a print in the collection of the late E. T. Roberts
Stockholders and Directors of the Pure Oil Company, 1903-1904. From *The Oil and Gas Journal*, August 23, 1934, p.77
Tarport and Tuna Valley, 1880. Mather Collection No. 1885
Ben Hogan's House in Tarport. From *The Life and Adventures of Ben Hogan*, facing p.208
Killing Time in Tarport. From *The Life and Adventures of Ben Hogan*, facing p.212
Richburg, New York, 1881. From A. C. Crum, *Romance of American Petroleum and Gas*, I, p.34
Acme Oil Company Refinery, Olean, New York. Mather Collection No. 1865
Map of the Western Pennsylvania Oil Region. From *Harper's Weekly*, August 26, 1882, p.538
Famous "Mystery Well"—No. 646, Cherry Grove, Warren County, 1882. Mather Collection No. 1631
Garfield, Cherry Grove, 1882. From *Harper's Weekly*, August 26, 1882, pp.536-537
"Hotel de Gunny Sack," Garfield, 1882. From *Harper's Weekly*, August 26, 1882, pp.536-537
Pumping Station, Garfield, 1882. From *Harper's Weekly*, August 26, 1882, pp.536-537
Famous Oil Scouts, Cherry Grove, 1882. From J. C. Tennent, *The Oil Scouts*, Frontispiece
Oil Scout Under Fire. From *Frank Leslie's Illustrated Newspaper*, April 4, 1885
A Torpedoed Well at Bradford. Mather Collection No. 1857
Phillips Well, Thorn Creek, Butler County, 1884. Mather Collection No. 1626
Christie Well, Thorn Creek, 1884. Mather Collection No. 1621
Oil Scouts at Phillips Well, Thorn Creek. Mather Collection Nos. 1078, 1627
Armstrong Well No. 2, Thorn Creek, 1884. Mather Collection No. 1617

CHAPTER X

Fire at the Grant Well. From *Frank Leslie's Illustrated Weekly*
General View, Acme Oil Company Refinery Ruins, 1880. Mather Collection No. 1724
Acme Refinery Ruins, 1880. Mather Collection No. 947
Acme Refinery Ruins, 1880. Mather Collection No. 893
Acme Refinery Ruins, South Perry Street Bridge, 1880. Mather Collection No. 1408
Fire and Flood, Titusville, 1892. Mather Collection No. 2054
Fire and Flood, Titusville, 1892. Mather Collection No. 2039

Fire and Flood, Titusville, 1892. Mather Collection No. 2053
Washington Street Ruins, Titusville, 1892. Mather Collection No. 2058
Fire and Flood Ruins, Titusville, 1892. Mather Collection No. 2055
Fire and Flood Ruins, Titusville, 1892. Mather Collection No. 2041
Mechanic Street Ruins, Titusville, 1892. Mather Collection No. 2057
Fire and Flood Refugees, Titusville, 1892. Mather Collection No. 2059
Seneca Street, Oil City, before Fire, 1892. From the collection of James H. Chickering of Oil City
Street Scene, Oil City, before Fire, 1892. From the collection of James H. Chickering
Street Scene, Oil City, before Fire, 1892. From A. C. Crum, *Romance of American Petroleum and Gas*, I, p.156
Centre Street Bridge, Oil City, 1892. From the collection of James H. Chickering
Centre Street, Oil City, 1892. From the collection of James H. Chickering
North Seneca Street, Oil City, 1892. From the collection of James H. Chickering
Fire and Flood Ruins, Titusville, 1892. Mather Collection No. 2052
Railroad Bridge, South Martin Street, Titusville, 1892. Mather Collection No. 2050
After the Fire and Flood, Oil City, 1892. From A. C. Crum, *Romance of American Petroleum and Gas*, I, p.157

CHAPTER XI

Drake Well, August 16, 1896. Mather Collection Nos. 462, 1324
Unveiling the Drake Monument, Woodlawn Cemetery, Titusville, October 4, 1901. Mather Collection No. 1337
Drake Monument, Woodlawn Cemetery. Drake Museum Collection Nos. 1348, 1347
Breaking Ground for the Drake Museum, October 3, 1911, Mather Collection No. 1315
Drake Museum, West Central Avenue, about 1915. Mather Collection No. 1170
E. C. Bell. Mather Collection No. 1873
Bell and Friends of the Museum. Mather Collection No. 1316
Bronze Plaque, Drake Well Monument
Drake Museum, Drake Museum Collection
Interior, Drake Museum. Drake Museum Collection
Drake Well Memorial Park near Titusville. Drake Museum Collection
Replica of the Drake Well, Drake Well Memorial Park, 1945. Drake Museum Collection

INDEX

Abbott, W. H., 69, 98, 102
Abbott & Harley, 70
Abbott & Harley Pipe Line Terminal, 69
Acme Oil Co., 93, 118, 125, 126, 127
Allegheny Transportation Co., 70
Allemagoozelum well, 31
Alvord House, 42
American Hotel, 36
American Petroleum Institute, 136, 141
American Revolution, 1
Anderson, George K., 35, 100, 102
Andover, 101
Andrews, C. H., 82
Andrews, F. W., 82
Andrews, W. C., 82
Andrews, Clark & Co., 104
Angell, Cyrus D., 105
Antwerp, 105
Appleton's Collegiate Institute, 98
Archbold, John D., 97, 101, 104
Armstrong, S. P., 124
Armstrong County, 103, 105, 106
Armstrong farm, 86
Armstrong well, 103
Armstrong well, No. 2, 124
Astor House, 61
Atkinson, 80
Atkinson farm, 76, 81, 82
Atlantic & Great Western Railroad, 37, 52

Backus, F. L., 82
Backus City, 80
bands, 27, 40, 102, 116
Barnsdall Oil Co., 96
Barnsdall well, 9
Barnum, P. T., 40
barrels, 15, 17, 54, 56, 77
Bartlett farm, 122
Bayonne, N.J., 115
Bear Creek, 109
Beardsley House, 82
Bell, E. C., 136, 139, 141
belt theory, 105
Benedict, W. B., 35
Bennett, Warner & Co., 93
Benninghoff, John, 24
Benninghoff farm, 21, 22, 23, 24
Benninghoff robbery, 24, 25
Benninghoff Run, 24, 25, 69, 70
Benson, Byron D., 115
"Benson's Folly," 115
Benton, J. G., 114, 115
Bethlehem, 136
Billings, Josh, 40
Bishop, Coleman E., 51
Bissell, George H., 4, 5, 46
Bissell, George H., & Co., 47
Blaney farm, 108
Blood farm, 29
Bloss, H. C., 40
Bloss, W. W., 40
boarding house, 45
boats, 53-57, 77
Bolivar, N.Y., 118
Bonanza Flats, 111
Bonta House, 63, 64, 98
Booth, John Wilkes, 61
Boston, Mass., 37
Boughton, H. H., Jr., & Co., 44
Boyd farm, 21-23
Boyle, Patrick C., 51
Boyle, Samuel, 118
Bradford, 64, 97, 103, 106, 109, 111-119, 121
Bradford & Foster Brook Railroad, 114
Bradford House, 113
Brewer, Francis Beattie, 4
Brewer, Watson & Co., 4, 6, 36
Brigham, Charles, 137
brokers, oil, 96, 105, 106
Brough, W., 35
Brown, Alex W., 88
Brown, Jack, well, 76, 80-81
Brown, John, 88
Brown, John F., 88
Brown, Samuel Q., 88
Brown Brothers Store, 88
Brown's Band, 40
Brown's Hotel, 45
Brundred refinery, 52
Buffalo, N.Y., 102
Bull, Ole, 91
Bullion, 64, 111
Bull Run, 18, 19
Butler, 103, 122
Butler County, 103, 105, 123

Cadwallader well, 119
Cady, D. H., 101
California, 82
Campbell farm, 76
Canadohta Chapter, D.A.R., 136, 140
Canadohta Lake, 93
Cappeau, J. P., 121
carbon oil, 1, 3
Carey, Nathaniel, 2
Carnegie, Andrew, 26
Carter, Charles Gibbs, 101
Carter, John J., 92, 101
Carter, Mrs. John J., 101
Centennial Exposition, 100, 110
Central House, 47
Central Petroleum Co., 46
Chase House, Pleasantville, 87
Cherry Grove, 119, 120, 121
Cherry Run, 34, 76
Cherrytree, 26
Cherrytree Run, 30, 31
Christie well, 103, 122, 123
Church Run, 103
churches, 98
"City of Stumps," 37
City Savings Bank, New Haven, Conn., 5
Clapp farm, 35
Clarendon, 119, 120
Clarion County, 103, 105
Clark, D. C., store, 45
Cleveland, Ohio, 14
"Coal Oil Johnny." *See* John W. Steele.
Cogswell, Mrs. J. H., 100
Columbia Conduit Co., 114
Columbia Cornet Band, 27
Columbia Oil Co., 26, 27
Commercial Hotel, 52
Commonwealth of Pennsylvania, 136, 141 142
Connecticut, 5
coopers, 17, 96
Corinthian Hall, 40
Cornplanter, 36
Cornplanter Run, 35
Corry, 37, 53, 58
Coryville, 115
Cottage Hill, 50
Coudersport, 36
Cow Run, 18
Crittenden House, 40
Crocker, Frederick, 43, 103
Crosby, Dixi, 4
Crossley well, 9
Cupler, Adam, Jr., 74

Dalzell Petroleum Co., 43
Danforth Hotel, 64, 65
danger from petroleum, 33
Dartmouth College, 4
Daughters of the American Revolution, 136 140
Davidson farm, 26
Delamater, George B., 15. *See also* Noble and Delamater.
Densmore, Amos, 66
derrick, inside of, 11
Dimick, George, 108, 119
Dimick, Nesbit & Co., 108
Dingley Run, 78
dippers, oil, 16
Dogtown, 105
Dom Pedro, Emperor of Brazil, 105
Downer, Samuel, 37
Downer Oil Works, 37
Drake, Edwin L., 1, 6, 7, 15, 36, 89, 98, 136, 138
Drake, John, 121
Drake Monument, Woodlawn Cemetery, 136, 137
Drake Museum, 138-139, 141-142
Drake Street School, 99
Drake well, 1, 7, 9, 36, 49, 76, 89, 125, 136, 137, 140-142
Drake Well Memorial Park, 140-142
Drake Well Monument, 140, 142
drama, oil, 33
drilling, 8, 10, 15, 86
Duncan & Co., 63

Eames, Charles J., 93
Eames Petroleum Iron Works, 93
East Titusville, 74
Economite Society, 76
Egbert, A. G., 26
Ellsworth Oil Well Association, 62
Emerson, E. O., 100
Emery, Lewis, Jr., 113, 116
Emery, Mrs. David, 136

Emperor of Brazil, 105
Empire Transportation Co., 67, 77, 110
Empire well, 9, 12, 43
engines and boilers, 8, 11, 28
Erie Hotel, 44
Erie Railroad, 37, 70
Eureka well, 63
Evans, Lewis, 1
Evans, Owen, 121
Eveleth, Jonathan G., 4, 5
Exchange Hotel, 77
exchanges, oil, 89, 97, 116, 119
excursions, 93

Fagundas, 74
Fairview, 110
Fairview Pipe Line Co., 108
Fanny Jane well, 108
Farmers Railroad, 52
Farnsworth, 120
Farrel, John, 102
Farrel, Nelson, 102
Farrel, Sadie, 102
Farrel farm, 14
Farwell, Annette, 8
Fee well, 76, 80, 81
fire and flood, 125-135
First National Bank, Titusville, 69
Fisher Brothers, 108
"fishing" for lost drilling tools, 16
Fisk, Jim, 91
flatboats, 55, 56
flatcar, horse-drawn, 25
Fletcher, R. D., 138
Foster, 102
Foster farm, 14, 19
Fountain well, 43
Foxburg, 103, 105, 106
Franklin, 2, 52, 61, 102, 105
Frazier well, 60
French Kate, 64, 107
Funk, A. B., 43
Funk farm, 12
Funkville, 43, 54

Garfield, 120
Garland, 53
George, T. M., 52
German Pietists, 76
Germany, 64
Gibbs, Russell & Sterrett, 102
Giles farm, 111
Gilmor City, 114
Goss, Marshall, 81
Goss farm, 76, 81
Gough, John B., 91
Grace, Peter, 119
Graff, Hasson & Co., 36
Grandins, 76
Granger & Co., 91
Grant, U.S., 91
Grant well, 62, 125
"Grasshopper City," 111
Great Western Run, 21
Greece, 108
Grumbine, Mrs. Samuel, 8
guide to oil region, 36

Hanover, N. H., 4
Hardison, W. L., 82
Harley, Henry, 69, 70, 91
Harley, Mrs. George R., 100
Harmonial well, 76, 85
Hayes, Rutherford B., 105
Heisman's Cooper Shop, 96
Henderson, J. M., Oil Brokerage Office, 96
Heron, A. M., 111
Heron, Dan, 121
Hess farm, 30, 31
Hibbard, "Pap," 36
Hinkley refinery, 40
historical materials on oil, 141-142
Hoffman House, 91
Hogan, Ben, 64, 107, 117
Hogback Hill, 51
Holmden farm, 60
Homestead well, 61
Hoskins Oil Co., 45
Hostetter, David, 110
"Hostetter's Stomach Bitters," 110
"Hotel de Gunny Sack," 120
"Hub of Oildom," 52
Hubbard, O. P., 4
Hughes, S., & Davis, 79
Hughes, S. B., 121
Hyde, Charles, 26
Hyde & Egbert farm, 26
Hyner farm, 61

Indian Rock Oil Co., 42
Irvineton, 70

Jackson, John, refinery, 93
James, Abram, 85
Jameson farm, 108
Jefferson City, 105
Jersey well, 26
Johnson, "Judge," 113
Jones, J. T., 116

Karns, Stephen Duncan, 109
Karns City, 103, 109
Kellogg, Clara Louise, 91
Kerr, James B., farm, 103
Keystone House, 52
kicking down a well, 10
Kier, Samuel M., 1, 3
King Bee well, 79
King of the Hills well, 25
Kirk, David, 116
Klinger farm, 74

Ladies' well, 66
Lady Brooks well, 20
Lady Stewart well, 82
lamps, oil, 141
Lee, J. W., 116
Limestone, N.Y., 103
Little, Levitt C., 112
Littleton, 112
Lizzie Toppling, 107
locating oil wells, 9
Logan, Olive, 40
long bridge, Oil City, 50

McClintock, Culbertson, 32
McClintock, Mrs. Culbertson, 32
McClintock, G. W., 21
McClintock farm, 46
McClintockville, 49
McClymonds farm, 109
McCray, James S., 26
McCray farm, 26
McElhenny Oil Co., 43
McKean County, 112
McKelvy, David K., 115
McKinney, J. C., 102
McKinney, John L., 102
McKinney Oil Office, 45
McMullen, J. C., 121
Magill wells, 19
Mansion House, 90
maps of the oil region, 1, 9, 36, 58, 115, 118
Marshall farm, 124
Mather, John A., v, 14, 41, 96, 136, 141
Meadville, 52, 61
Mehoopany well, 119
Methodists, 98
Michigan Rock Oil Co., 50
Miller farm, 16, 42, 66, 68, 70, 76
Millerstown, 103, 108, 109, 110
Mills, Major S. M., 25
Modoc, 108
Monitor refinery, 21
Moore House, 25
Moran House, 63
Moses, Job, 103
Mowbray, George M., 98
Muncy Station, 115
music, 17, 27, 30, 40, 91, 102, 116
Murphy, Michael, 116
Murphy well, 119
Mutual Pipe Line Co., 105
"Mystery Well"—No. 646, 119, 120

National wells, 88
New Brunswick Hotel, 65
New Hampshire, 112
New Haven, Conn., 5, 6
New York, 5
New York City, 1, 4, 32, 37, 66, 88, 115
New York & New Haven Railroad, 6
New York & Providence Petroleum Co., 88
Niagara & Pierson farm, 30
Niehaus, Charles Henry, 137
Nilsson, Christine, 91
Nitroglycerine explosion, 74
Noble, Orange, 15
Noble and Delamater well, 9, 14, 15, 70

Ocean well, 21
Octave Oil Co., refinery, 93
Oil City, 1, 35, 36, 46, 49, 50-53, 55, 70, 84, 97, 102, 103, 106, 108, 119, 125, 132-135
Oil City Daily Derrick, 51
Oil City Register, 36
Oil Creek Lake, 93
Oil Creek Railroad, 37, 42, 52, 53, 58, 59, 66, 68, 70, 89, 91
"Oil on the Brain," 17
Olean, N.Y., 118, 126

packet express boats, 57
Painesville, Ohio, 41
Parker's Landing, 64, 103, 106-109
Parshall Hotel, 90, 91, 102
"Peg Leg" Railroad, 114
Pennsylvania Historical and Museum Commission, 136
Pennsylvania Railroad, 37, 67, 110
Pennsylvania Rock Oil Co., Conn., 5
Pennsylvania Rock Oil Company, N.Y., 4